SNORKELING GUIDE TO MARINE LIFE

FLORIDA CARIBBEAN BAHAMAS

D1022715

Paul Humann • Ned DeLoach

NEW WORLD PUBLICATIONS, INC.

Printed by

Paramount Miller Graphics, Inc.

Jacksonville, Florida U.S.A.

Humann, Paul.
Snorkeler's Guide to Marine Life: Florida, Caribbean, Bahamas /
Paul Humann, Ned DeLoach.
p. cm.
ISBN: 1-878348-10-8

1. Marine biology–Florida. 2. Marine biology–Caribbean Sea. 3. Marine biology–Bahamas. 4. Skin diving–Florida. 5. Skin diving–Caribbean Sea. 6. Skin diving–Bahamas. I. DeLoach, Ned. II. Title.

QH105.F5H86 1995 574.93'759
QBI95-20086

CREDITS

Photographs by Paul Humann and Ned DeLoach
Art Director & Drawings: Michael O'Connell
Color Separations: Ben Vernon of Lithographic Services, Inc., Jacksonville, Fl; Buddy Waggoner of Paramount Miller Graphics, Jacksonville, Fl
Printed By: Paramount Miller Graphics, Jacksonville, FL
First Printing: 1995. Second Printing: 1998. Third Printing: 2001.
ISBN 1-878348-10-8 Library of Congress #95-067822
Copyright © 1995 New World Publications, Inc.
Copyright © 1995 Paul Humann
Copyright © 1995 Ned DeLoach
All rights reserved. No part of this book may be reproduced without prior written consent.
Publisher & Distributor: New World Publications, Inc.,1861 Cornell Rd., Jacksonville, FL 32207, (904) 737-6558

The Marine Wildlife Experience

Life began in the sea. Today, warm, shallow tropical waters still provide the quintessential environment for life on earth. Along Florida's southern shore, throughout the Bahamas' Island chain, and spreading south and west across the clear waters of the Caribbean Sea, is an underwater wilderness so rich in biological diversity that it must be seen to be believed.

Many inshore waters, only three to five feet deep, swirl with dazzling fish of every color and description. The spectacle includes a cornucopia of juveniles that spend their early lives in the shallows. Surrounding sands, corals and rocks house an unforgettable community of bizarre invertebrate life. If you love nature, this is where you must go.

Fortunately, little equipment, skill or training is necessary to visit these sunlit wonderlands. Hundreds of sites provide easy access and safe conditions for anyone with the desire to become an underwater explorer. Nearly all beach resorts and cruise ship lines provide fins, mask, safety vest and directions or guided tours to nearby snorkeling areas. It is not always necessary to find a close-in coral reef to discover marine life while snorkeling. Many species can be found on sand flats, in grass beds, along rocky shorelines, around dock pilings and at the edges of mangrove forests.

Snorkeling Guide to Marine Life is the first comprehensive, photographic identification reference designed specifically for the snorkeling naturalists. The 260 displayed species all inhabit waters from 15-foot depths to shore. This feature, combined with the book's user-friendly format, makes the quick, accurate identification of shallow-water species possible for the first-time snorkeler as well as veterans of many underwater outings.

Marine life identification adds an exciting dimension to snorkeling. All you need to get started is this book and a slate to jot down what you see. Underwater ID cards, displaying common fish and invertebrates, are also a great teaching tool. Then the hunt is on. The first step is to make yourself slow down and concentrate attention on one species at a time. The best way to discipline such behavior is by taking notes on an underwater slate.

After only a few hours in the water your personal sighting count will grow to 30, 40 species or more, but the real fun begins when an unfamiliar life form is sighted. Suddenly you are an underwater detective, looking for any clue that might reveal the identity of your mystery species. Everything depends on focus, stalking skills, attention to details and memory. Size, body shape, color, markings: any or all might be keys that could substantiate an ID and add another name to your growing list of species sightings.

Soon you will be setting personal goals and searching for a new species that captures your imagination. But be forewarned — marine life identification is addictive; your first snorkel could easily be the beginning of an adventure that lasts a lifetime. Most significant, however, is that, with every discovery, your commitment to preserving our fragile marine environment is awakened anew.

Contents

6. Swim With Pectoral Fins/Obvious Scales 34-43

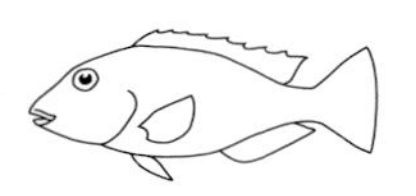
Parrotfish

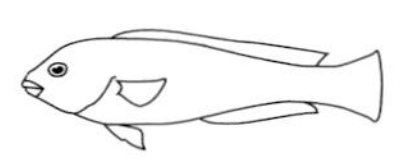
Wrasse

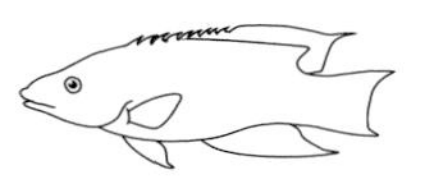
Hogfish/Wrasse

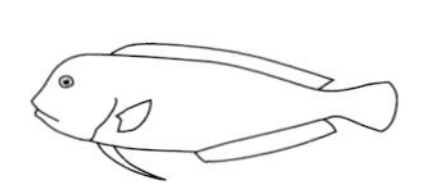
Razorfish/Wrasse

7. Reddish/Big Eyes 44-45

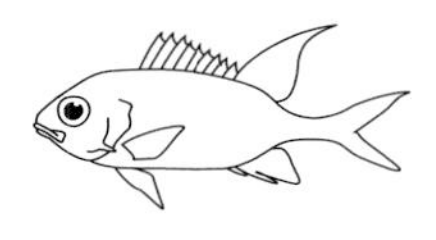
Squirrelfish

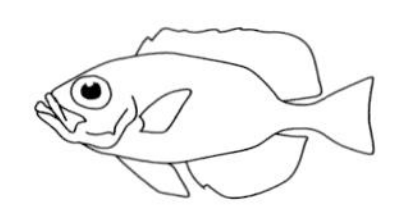
Bigeye

Cardinalfish

8. Bottom-Dwellers 46-51

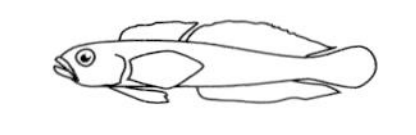
Goby

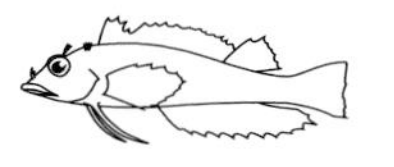
Blenny

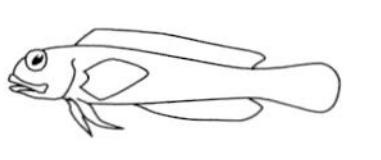
Jawfish

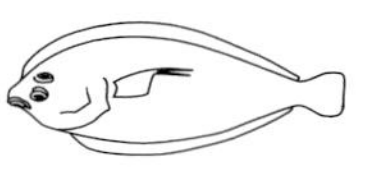
Flounder

Lizardfish

Flying Gurnard

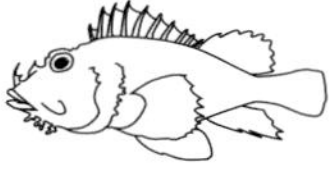
Scorpionfish

Frogfish

Seahorse

9. Odd-Shaped Swimmers 52-57

Trumpetfish

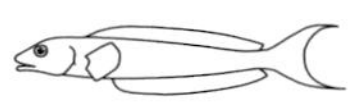
Tilefish

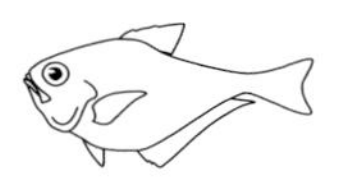
Sweeper

Goatfish

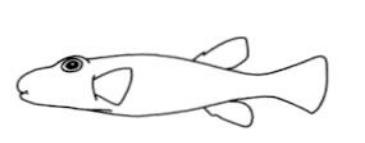
Smooth Puffer

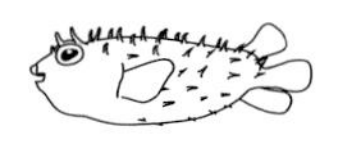
Spiny Puffer

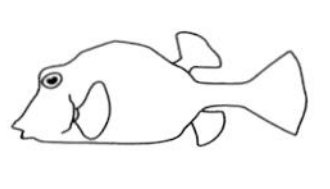
Boxfish

Triggerfish

Filefish

Drum

10. Eels, Rays & Sharks 58-61

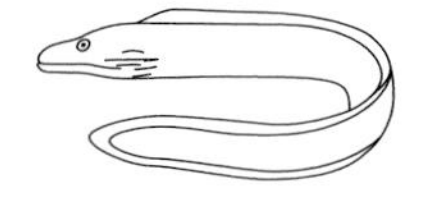
Moray

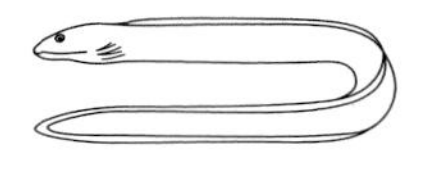
Snake Eel

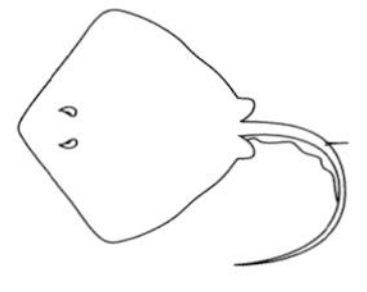
Ray

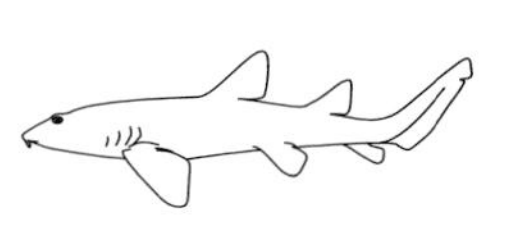
Nurse Shark

Branching & Pillar Corals

Mound & Boulder Corals

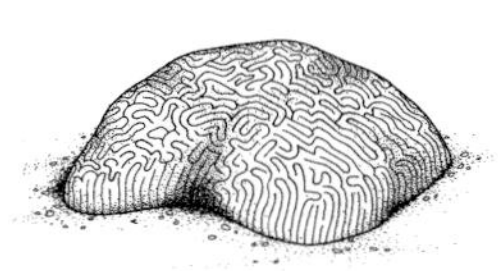
Brain Corals

Typical Shapes of Fire Corals
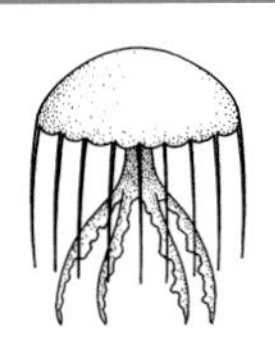

Hydroid

Jellyfish
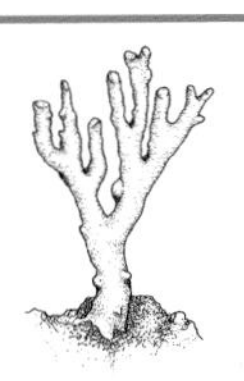

Sea Rod

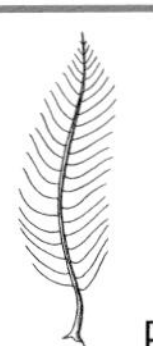
Sea Plume

Sea Fan

Anemone
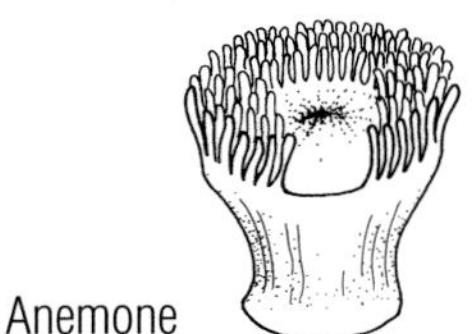

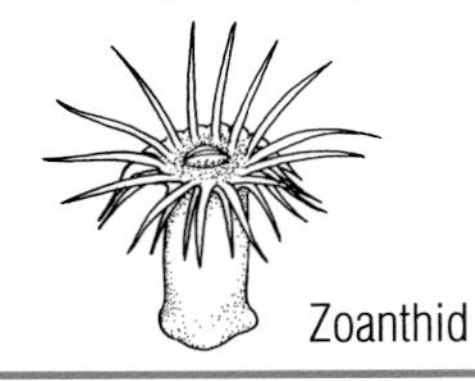
Zoanthid

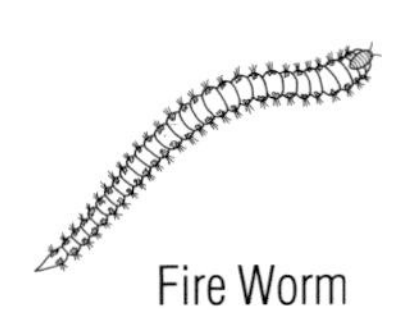
Fire Worm

Fan Worm

Tube Worms

Spaghetti Worm

Sponges 70-71

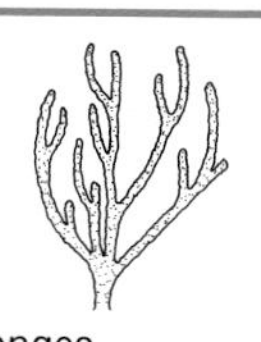

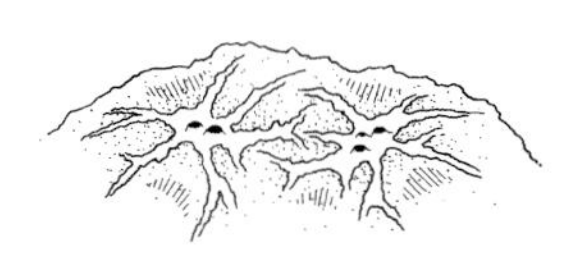

Typical Shapes of Sponges

Crustaceans 72-73

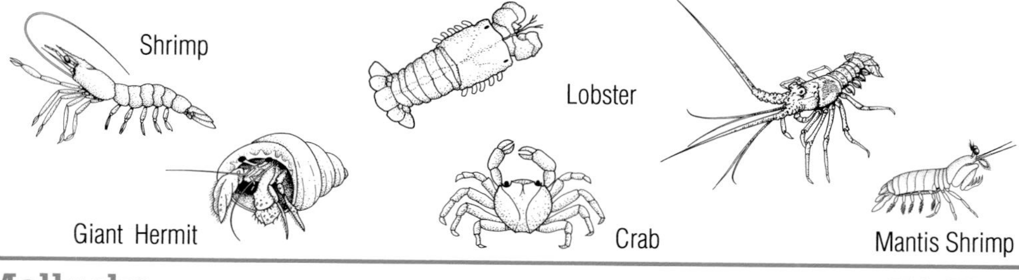

Mollusks 74-75

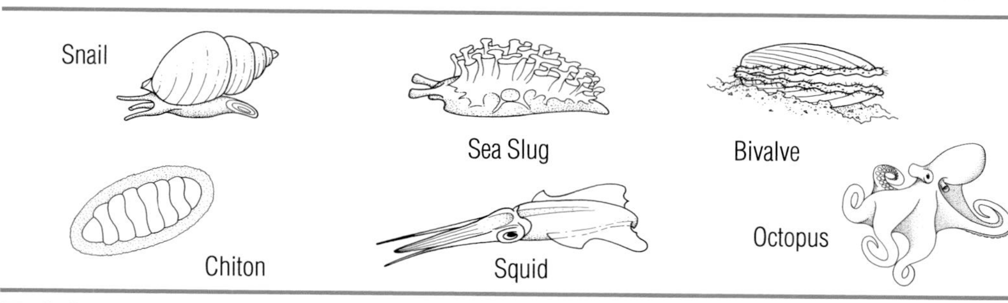

Echinoderms 76-77

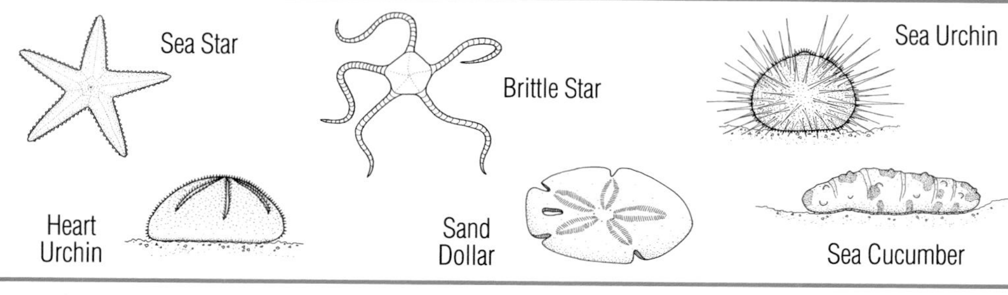

Marine Plants 78-80

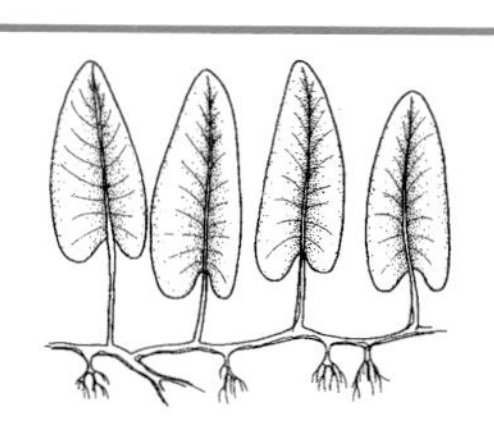

Typical Plant Shapes

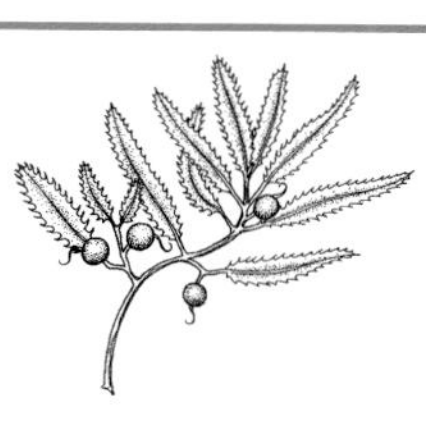

How To Use This Book

This comprehensive visual identification guide to the marine life inhabiting coastal waters of Florida, the Caribbean and Bahamas is designed for quick-reference. The nearly 260 included species are arranged in visually similar groupings. These include ten Fish Identification Groups and twelve Coral, Invertebrate and Marine Plant Identification Groups. Family members that make up a specific group are displayed by profile drawings on the four preceding Content pages. Families are scientific groupings based on evolutionary sequence, and, consequently, typically have similar physical characteristics. The Content pages provide, at a glance, a guide to where a sighted species can be found in the book. If a species' group is known by name, the reader can refer to the Fish Index on the inside front cover, or the Coral, Plant, Invertebrate Index on the inside back cover. An introduction of a family's behavioral and physical characteristics (that are observable by snorkelers) is presented at the beginning of each ID Group.

The information next to a photograph begins with a species' common name (that used by the general public). These reflect preferred names published by the American Fisheries Society. Below the common name, in italics, is the two-part scientific name. The first word (always capitalized) is the genus. The second word is the species. A species includes only animals that are sexually compatible and produce fertile offspring.

The profile drawing to the left of fish photographs shows distinctive features that differentiate that particular species from other similar appearing fish. Numbered arrows point out features explained in the bold text above the drawing. Terms used to describe markings and physical features are limited to those illustrated by the drawing on page 11. If additional information is included, it is presented in standard type. Size refers to the general size range of the fish that snorkelers are most likely to observe, followed by the maximum recorded size. Several species are exhibited in more than one photograph. This is necessary to demonstrate variations in color, markings and shape that occur within the species. Juvenile fish are often included because shallow inshore habitats are nursery grounds for many species.

All species in this book inhabit waters from 15 feet to shoreline although most also are found deeper. Not all, however, are found on coral reef habitats. Many inhabit open water, sand, surf zones, rocky shores, mangroves or manmade structures, such as dock pilings. Likewise, not all species inhabit waters throughout the region. For example, some might be specific to South Florida or the southern Caribbean.

Coral Reef Etiquette

Although stony corals appear to be non-living, they are actually colonies of tiny polyp animals that are easily killed or damaged by excessive contact. It is extremely important that snorkelers never stand on coral. Branching type corals, which form coral gardens in shallow areas, easily snap with pressure. Even an errant fin stroke or incidental contact from investigating hands can permanently damage the delicate outer covering of animals that form these magnificent sea sculptures.

It is equally important that nothing is taken from the sea. In the past, an uninformed public thought it acceptable to return from a snorkel with souvenirs including coral, sea fans and shells that, in many cases, take years to evolve.

Fish feeding has caused a nuisance at many popular snorkeling sites. Opportunistic feeders such as Sergeant Majors and Yellowtail Snapper have become conditioned to "Wonder Bread Lust". When a snorkeler, with or without handouts, enters the water, fish aggressively flock around. This unnatural activity corrupts the ecosystem's natural cadence.

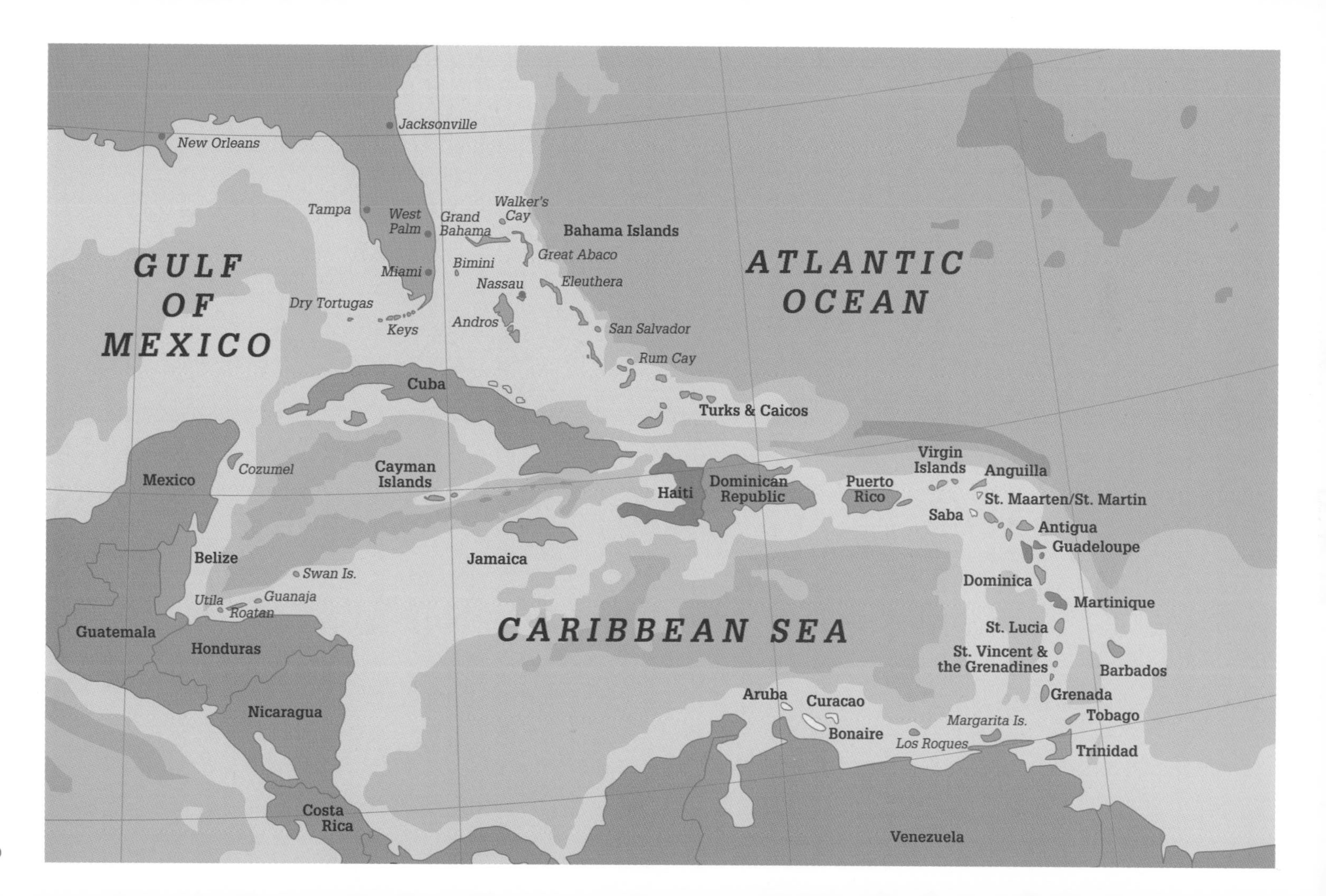
Jacksonville
New Orleans
Tampa
West Palm
Grand Bahama
Walker's Cay
Bahama Islands
Great Abaco
GULF OF MEXICO
ATLANTIC OCEAN
Bimini
Miami
Nassau
Eleuthera
Dry Tortugas
Andros
Keys
San Salvador
Rum Cay
Cuba
Turks & Caicos
Virgin Islands
Anguilla
Mexico
Cozumel
Cayman Islands
Haiti
Dominican Republic
Puerto Rico
St. Maarten/St. Martin
Saba
Antigua
Guadeloupe
Belize
Jamaica
Swan Is.
Dominica
Utila
Guanaja
Roatan
Martinique
Guatemala
Honduras
CARIBBEAN SEA
St. Lucia
St. Vincent & the Grenadines
Barbados
Aruba
Curacao
Grenada
Nicaragua
Bonaire
Margarita Is.
Tobago
Los Roques
Trinidad
Costa Rica
Venezuela

Marine Life Cautions

Contrary to popular belief there are few threats to the snorkeler from sea creatures. The reverse is generally the case, when uncaring snorkelers make contact with or molest marine life. Aggression toward humans is extremely rare on the reef. The few injuries that do occur are caused by contact with an animal's natural defense apparatus.

Of course, every beginning snorkeler wants to know about **sharks** and what to do if one is encountered. Unfortunately, these dramatic creatures are seldom if ever seen in snorkel areas. When they happen to venture near shore or reef during the day, they immediately depart at the sight of humans. If a shark is sighted, and it makes you feel uncomfortable, calmly leave the water. Bottom dwelling Nurse Sharks (p. 61), however, are sighted on a regular basis.

Possibly nothing is more unnerving to a beginning snorkeler than the approach of a **barracuda** (p. 17). These large silvery predators, which can reach lengths approaching six feet, are equipped with a formidable array of sharp teeth that they regularly display by slowly opening and closing their mouths. This is not an aggressive display; the fish are simply pumping water past their gills to aid respiration when not on the move. Although the fish certainly appear menacing, their habit of approaching snorkelers and following them about seems to be nothing more than curiosity. There are no substantiated attacks on snorkelers or divers unless spearfishing was involved. So relax and enjoy the company of these magnificent creatures.

Another sea creature with an undeserved reputation is the **moray** (p. 58). Like the barracuda, morays also have the unnerving habit of opening and closing their mouths to aid respiration. Although they can deliver a nasty bite, this behavior is rare. Generally, morays occupy crevices in the reef with only their heads exposed. However, those conditioned to hand-feeding often leave their lairs to greet approaching snorkelers.

Stingrays (p. 60) are flattened fish that spend most of their time lying on the bottom. Rays often flip sand over their backs to conceal their location. A venomous barb on the base of their tails can inflict a severe wound, but is never used aggressively. This defense only comes into play, as a reflex, if the animal is stepped on or handled.

Another venomous bottom dweller is the **scorpionfish** (p. 51). These strange looking fish rely on camouflage to blend in with the background. From concealed positions these ambush predators wait patiently for unsuspecting fish to swim close to their cavernous mouths — then suddenly, gulp! Their pain-producing toxin is injected from the front dorsal spines.

Three to six inch long **fireworms** (p. 69), often seen slowly crawling about, get their name from the multitude of tiny venomous bristles along their sides. If touched, the bristles easily break off in the skin, causing a painful burning sensation and irritating wound.

Long-spined Urchins (p. 77) should be avoided. These algae eating echinoderms inhabit many shallow-water areas. During the day most seek the shelter of reef recesses with only their spines exposed. Their sharp spines easily penetrate skin and break. Spines are not venomous but impossible to remove; however, they will dissolve within a few days. The affected area should be treated for infection.

Invertebrates from the phylum cnidaria (nigh-DARE-ee-uh), which include corals, hydroids, jellyfish, anemones and others, share the unique characteristic of bearing thousands of stinging capsules, called nematocysts. These minute capsules, located primarily on the tentacles, are used for both capturing prey and defense.

The stings of most cnidarians have no harmful effect on snorkelers, but a few are quite toxic and should be avoided. The sting of nearly all stony corals, gorgonians and anemones are not felt by humans, but any contact causes a degree of harm to the creature. However, contact with **hydroids** (p. 65), some **jellyfish** (p. 65), and **fire coral** (p. 64) can cause mild to severe pain. In the event of a sting, never rub the affected area or wash with water or soap. Both actions can cause additional nematocysts to discharge. Saturating the area with vinegar will immobilize unspent capsules; a sprinkling of meat tenderizer may help alleviate the symptoms.

Anatomy and Description

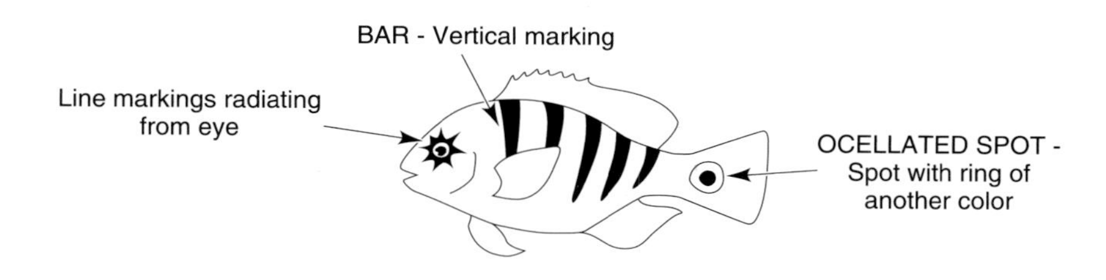

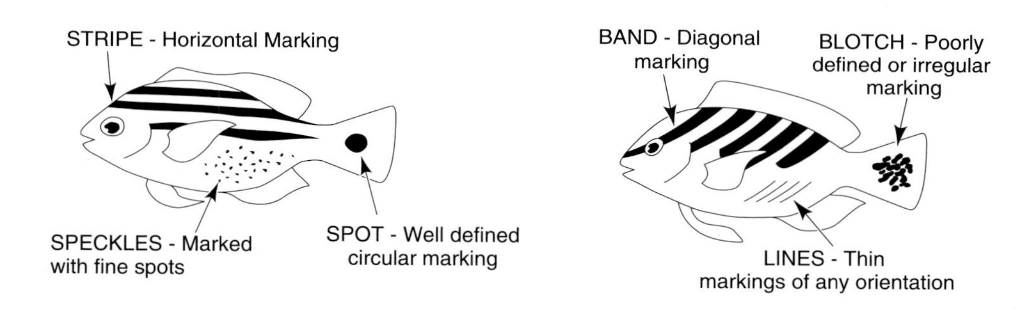

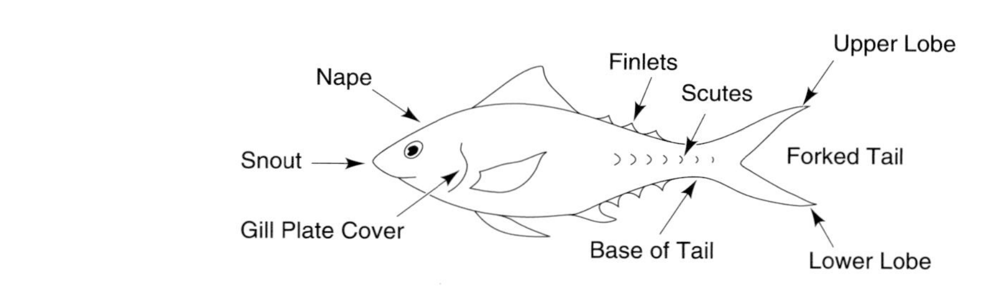

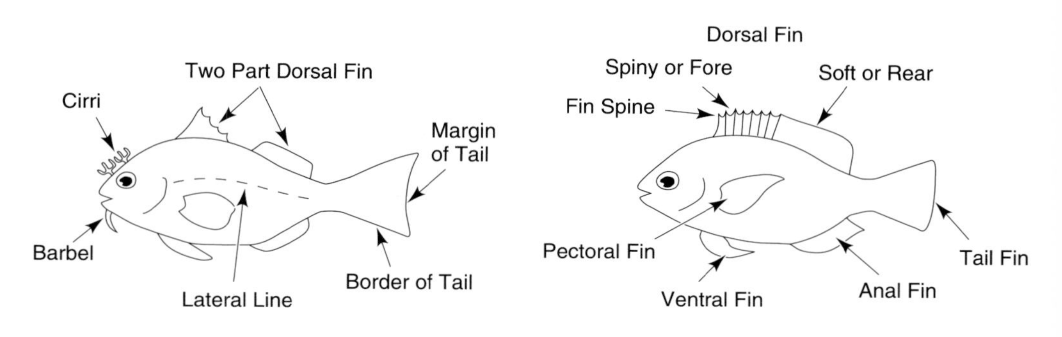

FISH IDENTIFICATION GROUP 1

Disks & Ovals/Colorful

Butterflyfish– Angelfish – Surgeonfish

This ID Group consists of fish that are thin-bodied and have round or oval profiles. All have small mouths and are generally quite colorful.

BUTTERFLYFISH - Chaetodontidae - These round, thin-bodied fish are easy to recognize as they flit about the reefs in search of food. They travel alone or in pairs, using keen eyesight to spot tiny worms or exposed polyps. Their small size (usually less than six inches) and slightly concave heads make them easy to distinguish from the larger, similarly shaped angelfish which have rounded foreheads. They are silver to white, with yellow tints and dark markings. Their eyes are concealed by dark bands on the head.

ANGELFISH - Pomacanthidae - Adult angelfish swim gracefully and generally grow to more than a foot in length. They have long dorsal and anal fins, and rounded foreheads. Juveniles differ significantly in both markings and color, and are difficult to identify to species.

SURGEONFISH - Acanthuridae - A spine as sharp as a surgeon's scalpel, located on each side of the body at the base of the tail, is the origin of this family's common name. When not in use for defense, their spines fold forward against the body. All three species of surgeonfish are common reef inhabitants that often mix in loose aggregations as they move about the reef feeding on algae.

BANDED BUTTERFLYFISH

Chaetodon striatus

SIZE: 3-5 in., max. 6 in.

1. Two wide, black midbody bands.

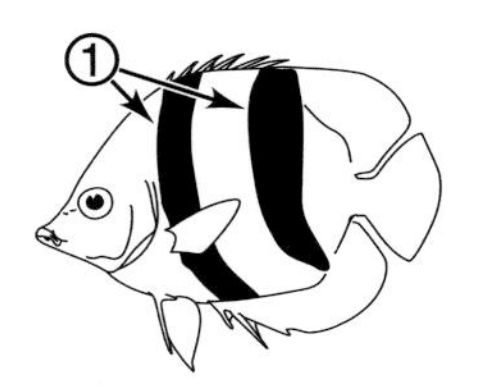

FOUREYE BUTTERFLYFISH

Chaetodon capistratus

SIZE: 3-5 in., max. 6 in.

1. Black spot, ringed in white, on rear body.

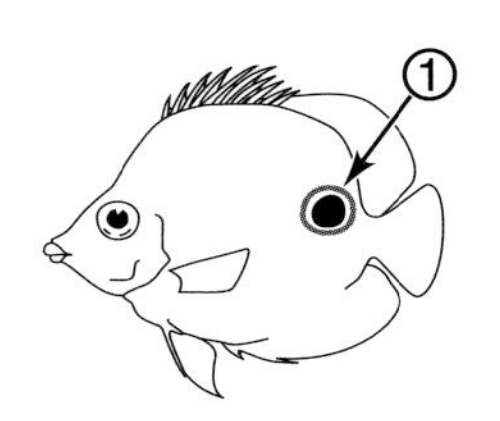

SPOTFIN BUTTERFLYFISH

Chaetodon ocellatus

SIZE: 3-6 in., max. 8 in.

1. Fins bright yellow. 2. Black dot on outer edge of rear dorsal fin.

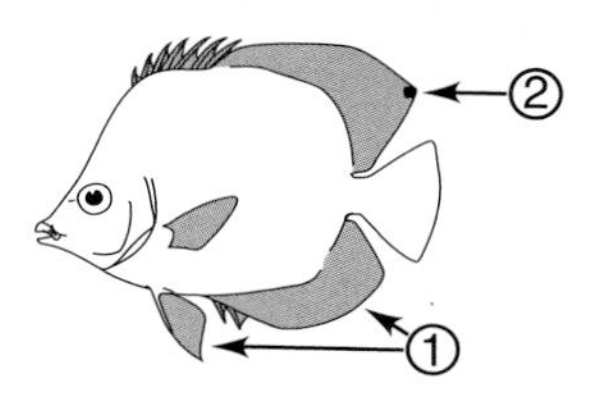

QUEEN ANGELFISH

Holacanthus ciliaris

SIZE: 8-14 in., max. 18 in.

1. Dark blue spot on forehead, speckled with blue, forms "crown." 2. Tail yellow.

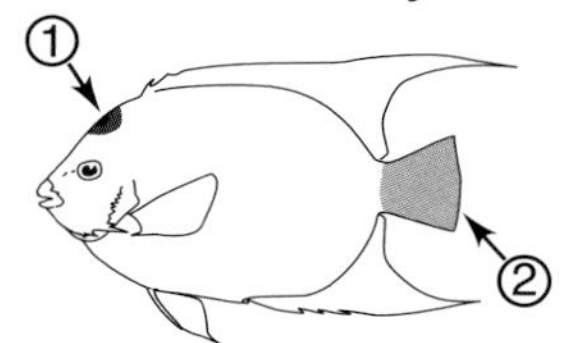

Queen Angelfish Juvenile

SIZE: 1-5 in.

1. Second blue body bar is curved.

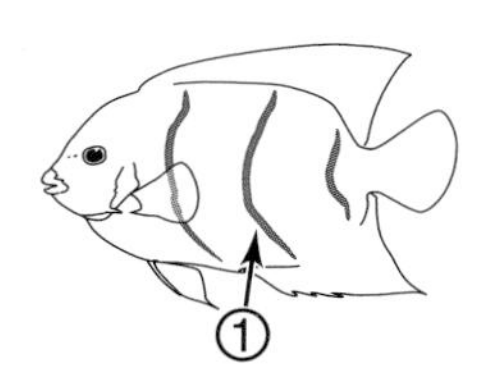

BLUE ANGELFISH

Holacanthus bermudensis

SIZE: 8-14 in., max. 18 in.

1. Tail and pectoral fins bordered in yellow.

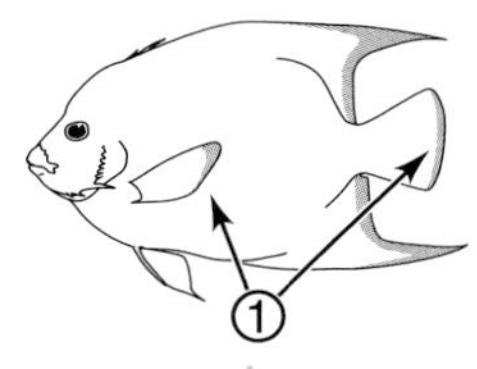

FRENCH ANGELFISH

Pomacanthus paru

SIZE: 10-14 in., max. 18 in.

1. Bright yellow rims on scales.

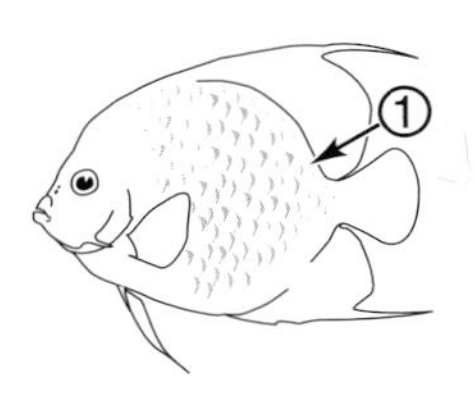

French Angelfish Juvenile

SIZE: 1-5 in.

1. Rounded tail with yellow border forms an oval.

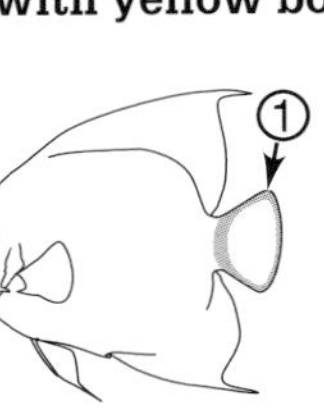

GRAY ANGELFISH

Pomacanthus arcuatus

SIZE: 10-18 in., max. 2 ft.

1.Yellow inner face of pectoral fin. 2. Square-cut tail.

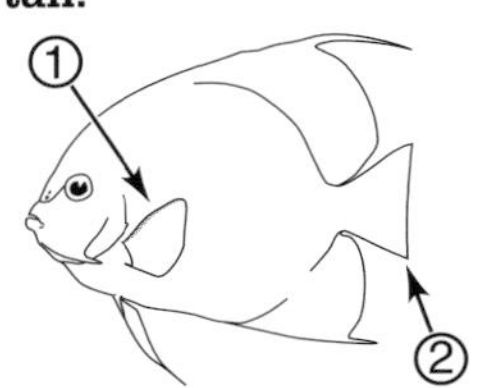

ROCK BEAUTY

Holacanthus tricolor

SIZE: 5-8 in., max. 12 in.

1. Yellow forebody and tail. 2. Mid and rear body black.

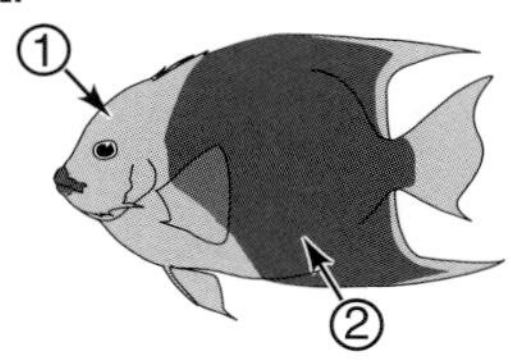

OCEAN SURGEONFISH

Acanthurus bahianus

SIZE: 6-12 in., max. 15 in.

Uniform color with no body bars.

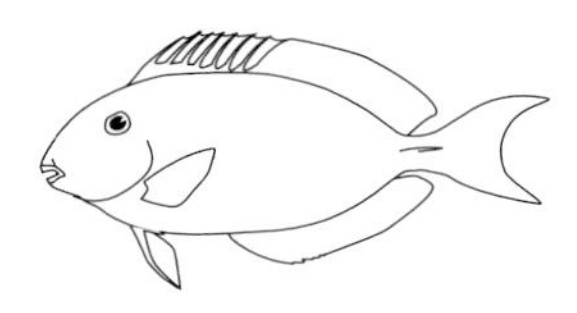

DOCTORFISH

Acanthurus chirurgus

SIZE: 6-12 in., max. 14 in.

1. Always have body bars.

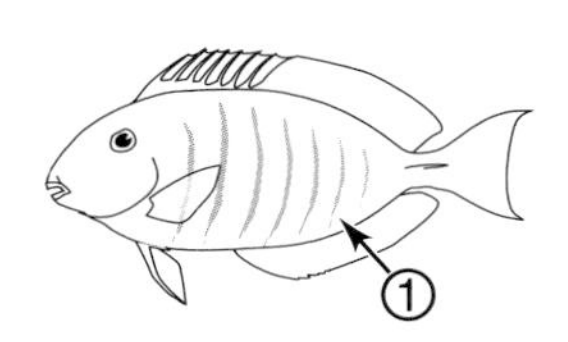

BLUE TANG

Acanthurus coeruleus

SIZE: 5-10 in., max. 15 in.

1. White or yellow spine on base of tail.

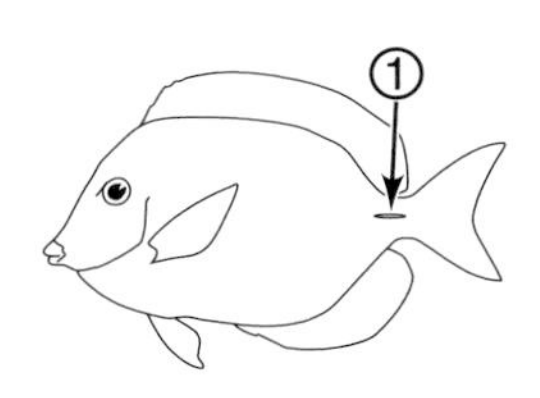

Blue Tang Juvenile

SIZE: 2-4 in.

1. Yellow tail. Very young bright yellow overall.

FISH IDENTIFICATION GROUP 2

Silvery

Jack - Porgy - Others

This ID Group consists of fish that are silver to gray in color, and are generally unpatterned; however, several species have bluish, yellowish or greenish tints and occasional markings. All have forked tails.

JACK - Carangidae - Jacks are strong-swimming predators of the open sea. Though schools occasionally pass over shallow reefs and through the surf zone in search of small fish and crustaceans, only a few species are seen in these areas on a regular basis.

PORGY - Sparidae - These generally silvery fish, with high back profiles and steep heads, are regularly sighted in snorkeling areas. They commonly display a yellow wash over their backs, and develop darkish barred patterns while feeding in the sand for shellfish and crabs. Although a bit tricky, most porgies can be identified to species.

NEEDLEFISH & HALFBEAK - Belonidae & Exocoetidae - Both needlefish and halfbeaks are long, thin silvery fish that generally swim just under the surface, often in shallow water habitats. The two families are easily separated; needlefish have upper and lower jaws of nearly equal length, while halfbeaks have short upper and long lower jaws. Visually distinguishing between the five species of needlefish and three halfbeaks that inhabit the region is quite difficult.

BAR JACK

Caranx ruber

SIZE: 8-14 in., max. 2 ft.

1. Black runs under dorsal fin and onto lower tail.

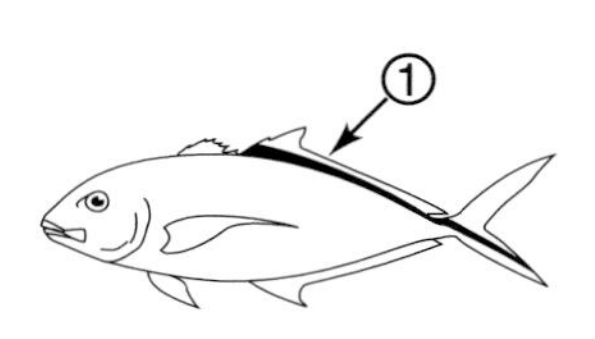

PALOMETA

Trachinotus goodei

SIZE: 7-14 in., max. 18 in.

1. Four dark body bars. 2. Extremely long dorsal and anal fins.

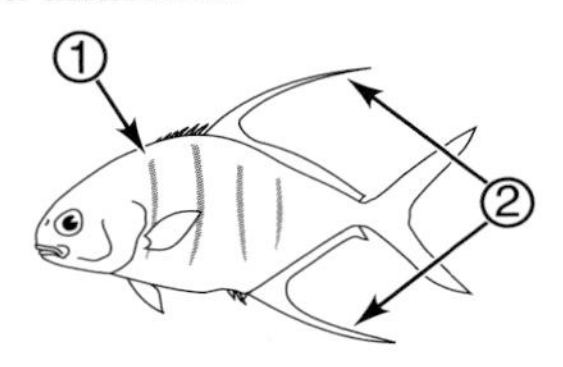

HOUNDFISH

Tylosurus crocodilus

SIZE: 2-3½ ft., max. 5 ft.

1. Black lateral keel on each side of tail base.

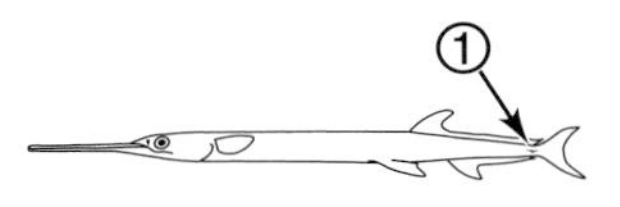

BALLYHOO

Hemiramphus brasiliensis

SIZE: 8-12 in., max. 16 in.

1. Upper lobe of tail yellow.

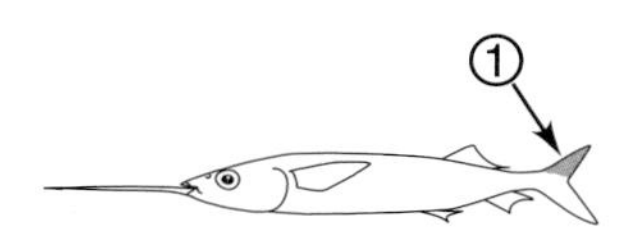

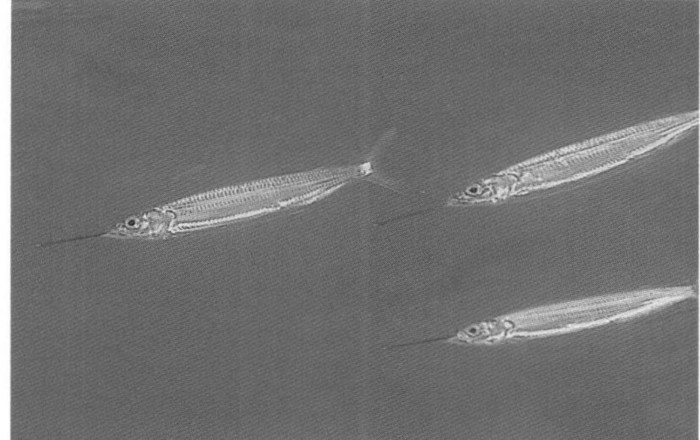

GREAT BARRACUDA

Sphyraena barracuda

SIZE: 1½-3 ft., max. 6 ft.

1. Large underslung jaw, pointed teeth often obvious.

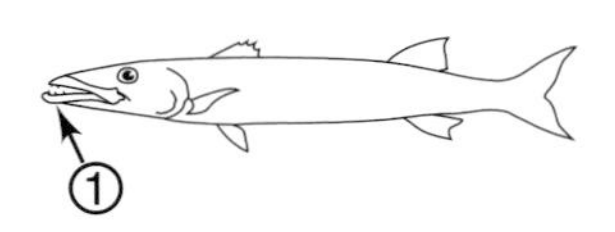

WHITE MULLET

Mugil curema

SIZE: 8-12 in., max. 15 in.

1. Large spot on base of pectoral fin. 2. Dusky margin on tail.

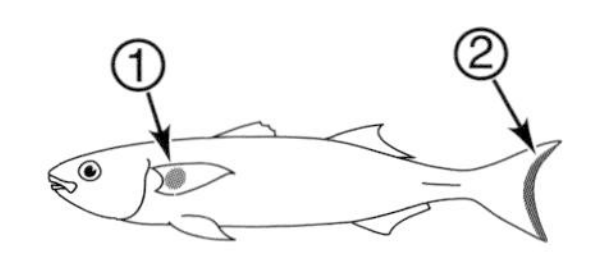

CHUB

Kyphosus sectatrix

SIZE: 1-2 ft., max. 2 1/2 ft.

"Football-shaped" body.

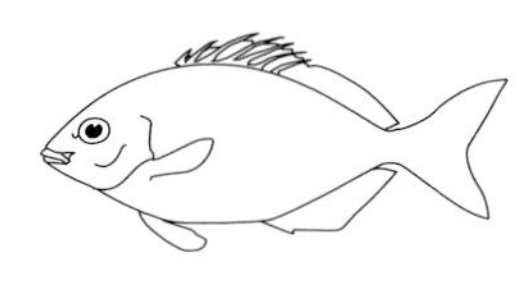

YELLOWFIN MOJARRA

Gerres cinereus

SIZE: 8-12 in., max. 16 in.

1. Yellow ventral fins. 2. Several indistinct bars on body.

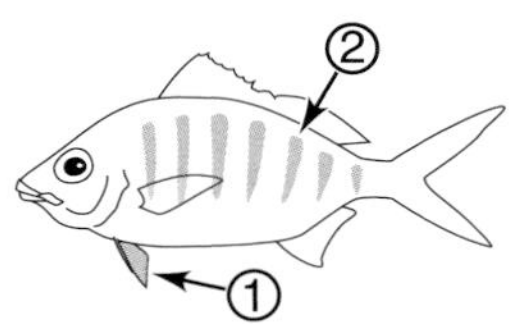

SLENDER MOJARRA

Eucinostomus jonesi

SIZE: 4-6 in., max. 8 in.

Slender body.

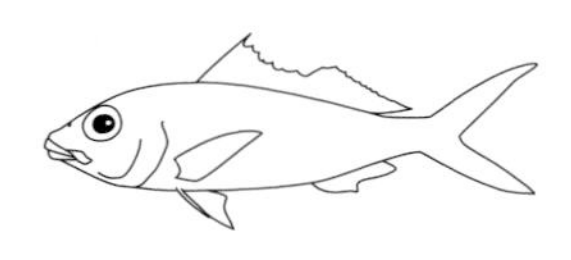

SEA BREAM

Archosargus rhomboidalis

SIZE: 5-8 in., max. 13 in.

1. Dusky spot behind gill cover.

SILVER PORGY

Diplodus argenteus

SIZE: 4-8 in., max. 12 in.

1. Black spot on tail base.

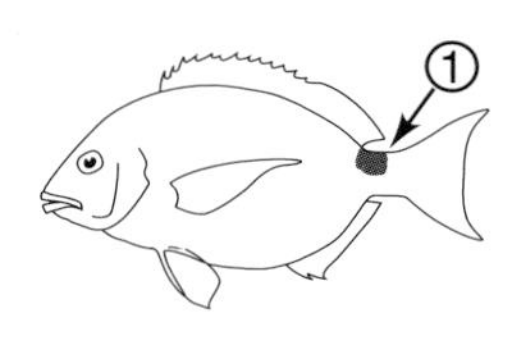

SAUCEREYE PORGY

Calamus calamus

SIZE: 8-14 in., max. 16 in.

1. Bluish saucer-shaped line below eye. Often yellow wash across back.

JOLTHEAD PORGY

Calamus bajonado

SIZE: 1-1½ ft., max. 2 ft.

1. High back profile. 2. Large lower jaw with thick lips.

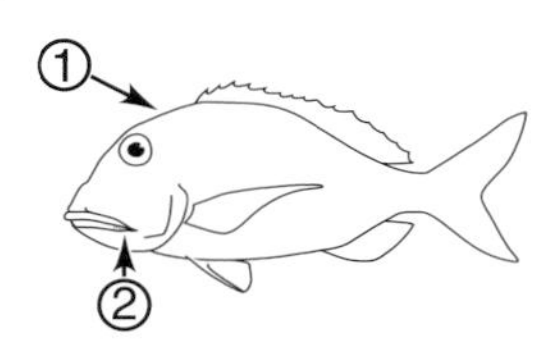

PLUMA

Calamus pennatula

SIZE: 8-12 in., max. 15 in.

1. Blue area behind eye. 2. Irregular blue lines below eye.

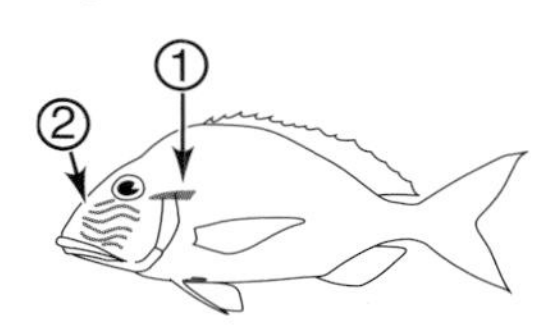

FISH IDENTIFICATION GROUP 3

Sloping Head/Tapered Body

Grunt - Snapper

This ID Group consists of fish that have what can best be described as a basic, "fish-like" shape, relatively large mouths and notched tails.

GRUNT - Haemulidae - Grunts are closely related to snapper, but are generally smaller, with more deeply notched tails. They also lack the snappers' sharp canine teeth. Most are colorfully striped. During the day grunts congregate in small groups to large schools that drift in the shadows of reefs. At night, the nocturnal feeders scavenge the sand flats and grass beds near reefs for crustaceans. Grunts often make up the largest biomass on reefs in continental or insular shelf areas that have large expanses of grass beds and sand flats. Grunt populations are less prominent around islands lacking these habitats. Adults all have distinctive features and are fairly easy to distinguish. Early juveniles (one to two inch), which commonly inhabit snorkeling areas, are quite similar in appearance and, therefore, difficult to identify to species.

SNAPPER - Lutjanidae - The behavior of snapping their jaws when hooked gives snappers their name. They are medium-sized (usually one to two feet), oblong shaped fish with triangular heads. All have a single, continuous dorsal fin that is often high in the front, and shallow, notched tails. They have slightly upturned snouts, large mouths, and prominent canine teeth near the front of the jaw. Larger snapper tend to be solitary: Mahogany, Gray and Lane Snappers often gather in small groups; Yellowtail Snappers swim in loose aggregations well off the bottom; Schoolmasters often form large schools. The popular food fish known as Red Snapper are deep water species.

FRENCH GRUNT

Haemulon flavolineatum

SIZE: 6-10 in., max. 1 ft.

1. Yellow stripes below lateral line set on diagonal.

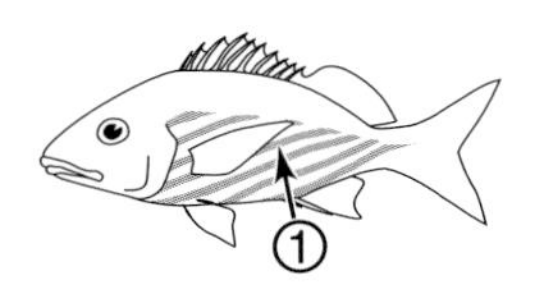

French Grunt Juvenile

SIZE: 2-4 in.

Juvenile grunts are quite similar in appearance. Most resemble this young French with two to three body stripes and a spot on tail base.

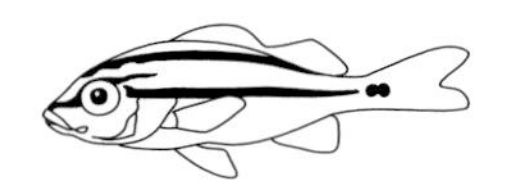

SMALLMOUTH GRUNT

Haemulon chrysargyreum

SIZE: 7-9 in., max. 10 in.

1. Five or six yellow stripes. 2. Fins yellow.

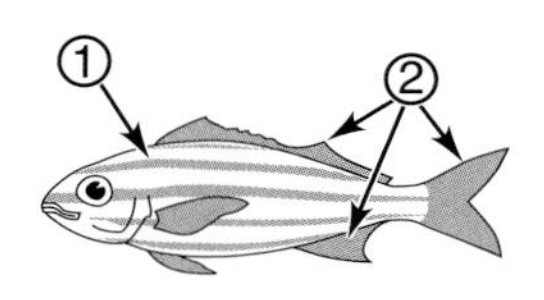

WHITE GRUNT

Haemulon plumieri

SIZE: 8-14 in., max. 18 in.

1. Stripes only on head.

BLUESTRIPED GRUNT

Haemulon sciurus

SIZE: 8-14 in., max. 18 in.

1. Dark tail and rear dorsal fin. 2. Blue stripes.

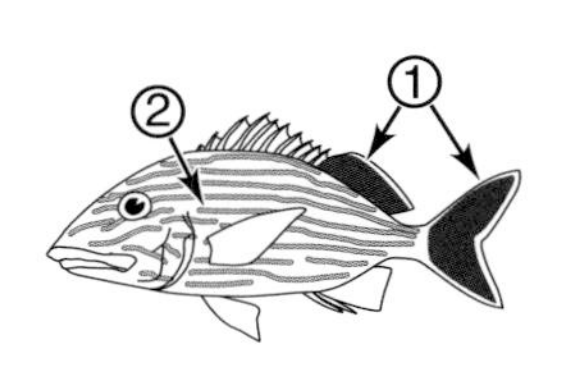

CAESAR GRUNT

Haemulon carbonarium

SIZE: 7-12 in., max. 15 in.

1. Yellow to bronze stripes. 2. Dusky rear dorsal, anal and tail fins.

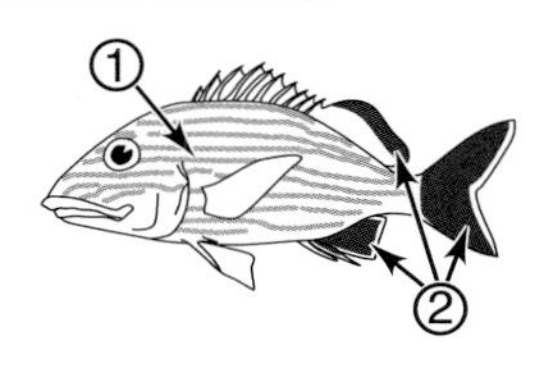

TOMTATE

Haemulon aurolineatum

SIZE: 5 - 8 in., max. 10 in.

1. Yellow stripe runs from snout to tail. 2. Dark spot on base of tail (not always present).

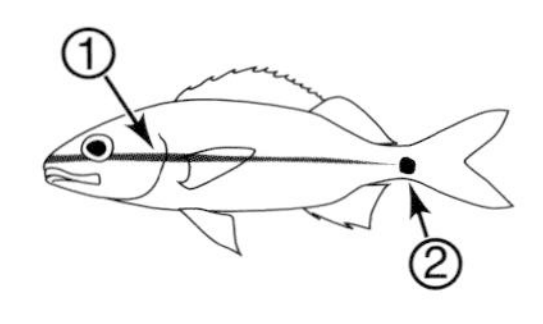

SAILORS CHOICE

Haemulon parra

SIZE: 8 - 12 in., max. 17 in.

1. Fins dusky to dark.

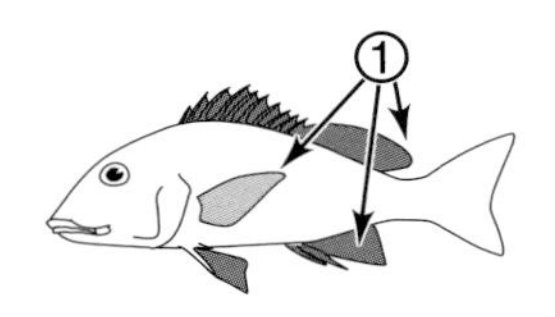

MARGATE

Haemulon album

SIZE: 10 - 20 in., max. 27 in.

1. High back profile. Largest grunt.

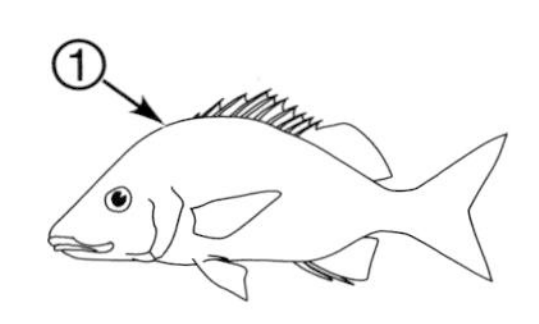

PORKFISH

Anisotremus virginicus

SIZE: 6 - 10 in., max. 14 in.

1. Two black bands on head. 2. High back profile.

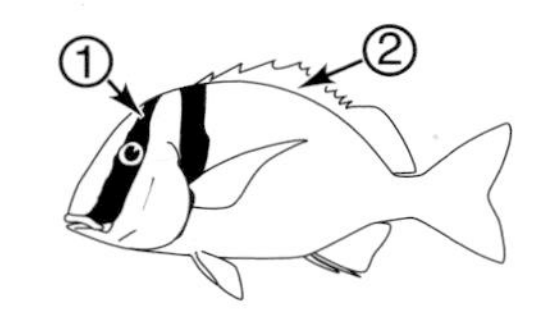

GRAY SNAPPER

Lutjanus griseus

SIZE: 10 - 18 in., max. 2 ft.

Gray without markings. **1. In shallow water often display darkish band through eye.**

SCHOOLMASTER

Lutjanus apodus

SIZE: 8 - 18 in., max. 2 ft.

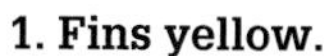

1. Fins yellow.

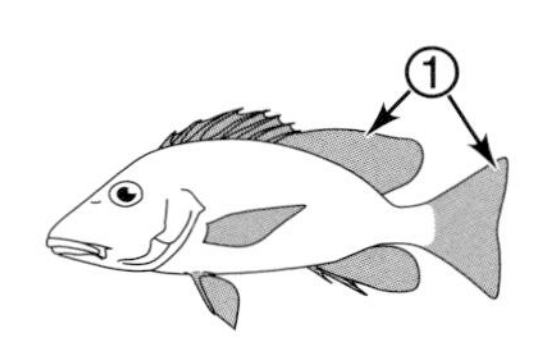

MAHOGANY SNAPPER

Lutjanus mahogoni

SIZE: 7 - 12 in., max. 15 in.

1. Reddish margin on tail.

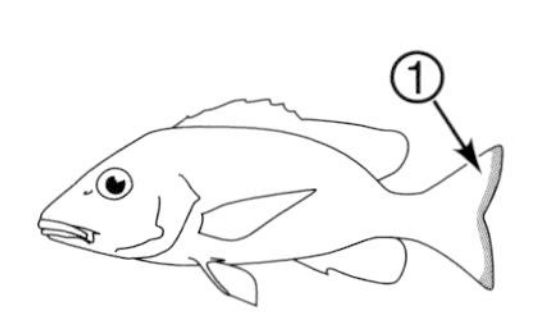

YELLOWTAIL SNAPPER

Ocyurus chrysurus

SIZE: 1 - 2 ft., max. 2 1/2 ft.

1. Yellow midbody stripe and tail.

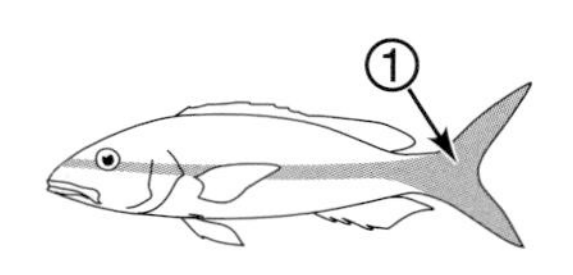

FISH IDENTIFICATION GROUP 4

Small Ovals

Damselfish - Chromis/Damselfish - Hamlet/Seabass

This ID Group consists of small fish with oval profiles.

DAMSELFISH - Pomacentridae - These energetic little fish are an evident part of the coral reef community; although most of the adults are drab, the juveniles are quite colorful. Several species spend their days busily tending and patrolling a private algae patch that is pugnaciously defended from intruders. When the domain of the Cocoa, Dusky, Longfin or Three Spot is threatened, the fish dart back and forth with fins erect, ready to attack. Even snorkelers who approach too closely may receive a sharp nip from these feisty fish. Egg clusters are also defended in this manner by some males. The Yellowtail, Beaugregory and Bicolor Damselfish are territorial, but much less aggressive.

Distinguishing between the dark-bodied Longfin, Dusky, Beaugregory, Cocoa and Three Spot Damselfish is a bit confusing; however, subtle differences make underwater identification possible. Except for the two sergeants, juvenile damselfish differ dramatically from adults in both color and markings, and although similar, are easily distinguished.

CHROMIS/DAMSELFISH - Pomacentridae - Members of the genus *Chromis* are part of the damselfish family, but are discussed separately because the group carries its own common name, and its members are different in appearance and behavior. The Blue and Brown Chromis are the most frequently seen species. Both are somewhat elongated, and have deeply forked tails. They swim in small to large aggregations well above the reefs, feeding on plankton.

HAMLET/SEABASS - Serranidae - Hamlets of the genus *Hypoplectrus* are members of the seabass family. The hamlets' flatheaded profiles easily distinguish them from damselfish, which have rounded heads. Most are colorful, small (generally three to five inches), and have nearly identical body shapes; however, there are several distinctive color patterns and markings, generally making identification simple.

BICOLOR DAMSELFISH

Stegastes partitus

SIZE: 2 - 3 1/2 in., max. 4 in.

1. Forebody is usually black. Often very common. Stay close to bottom.

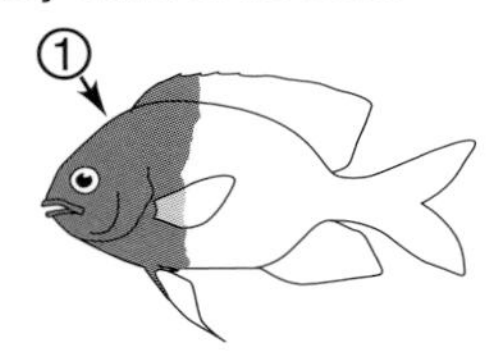

DUSKY DAMSELFISH

Stegastes fuscus

SIZE: 3-5 in., max. 6 in.

1. Dorsal and anal fins rounded and rarely extend beyond tail base.

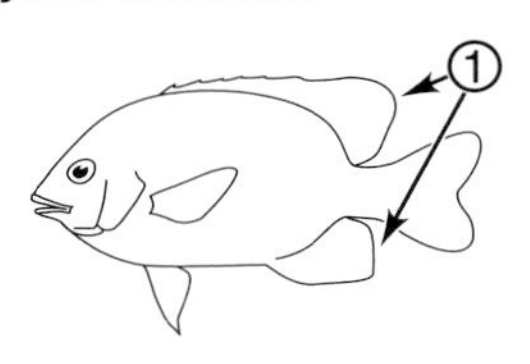

Dusky Damselfish Juvenile

SIZE: $1^1/_2$-$2^1/_2$ in.

1. Orange wash from snout to mid dorsal fin.

LONGFIN DAMSELFISH

Stegastes diencaeus

SIZE: 3-4 in., max. 5 in.

1. Dorsal and anal fins elongated and pointed.

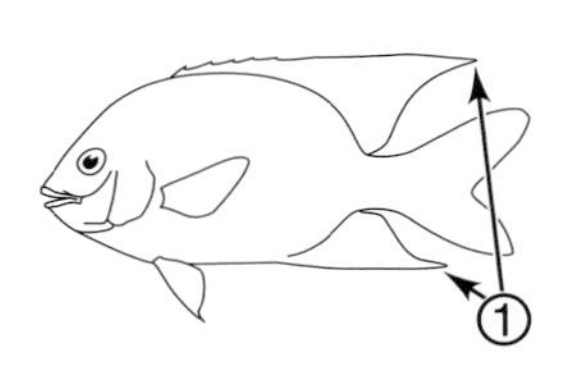

Longfin Damselfish Juvenile

SIZE: 2-3 in.

1. Brilliant blue lines run from snout and down back.

COCOA DAMSELFISH

Stegastes variabilis

SIZE: 3-4 in., max. 5 in.

1. Area around eyes dark. 2. Dark spot on upper tail base.

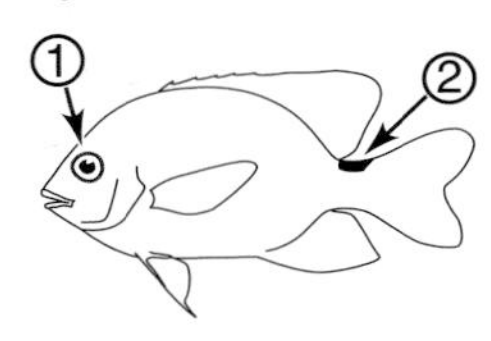

Cocoa Damselfish Juvenile

SIZE: $1\frac{1}{2}$-$2\frac{1}{2}$ in.

1. Blue wash from snout runs down back.
2. Spot on upper tail base.

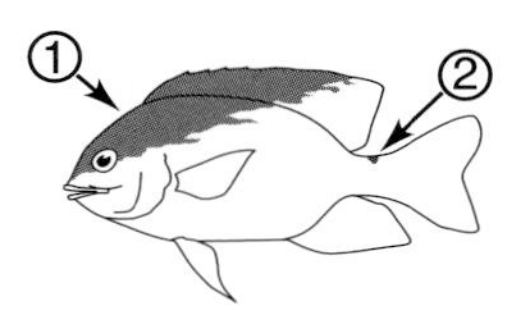

BEAUGREGORY

Stegastes leucostictus

SIZE: $2\frac{1}{2}$-$3\frac{1}{2}$ in., max. 4 in.

1. Pale to yellowish tail.

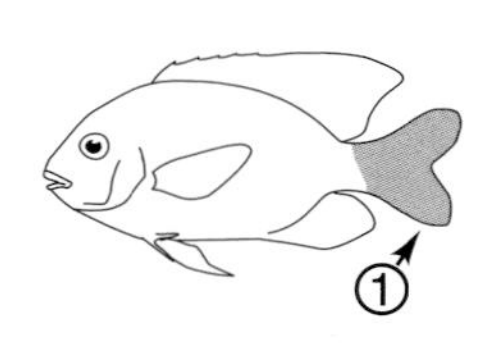

Beaugregory Juvenile

SIZE: $1\frac{1}{2}$-2 in.

1. Blue wash on back. No spot on tail base.

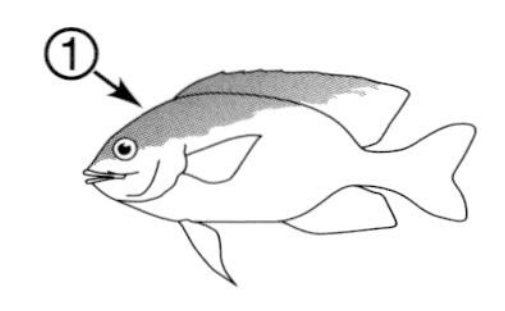

THREESPOT DAMSELFISH

Stegastes planifrons

SIZE: 3-4 in., max. 6 in.

1. Yellow crescent above eyes. 2. Spot on top of tail base and base of pectoral fin.

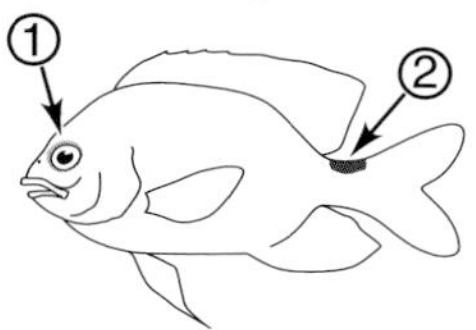

Threespot Damselfish Juvenile

SIZE: 1½-2½ in.

1. Black saddle on upper tail base. 2. Second black spot on back.

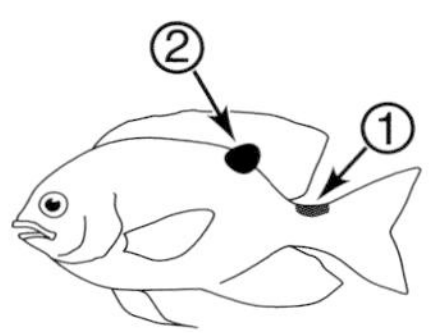

SERGEANT MAJOR

Abudefduf saxatilis

SIZE: 4-6 in., max. 7 in.

1. Five black body bars. Male becomes blueish during mating season.

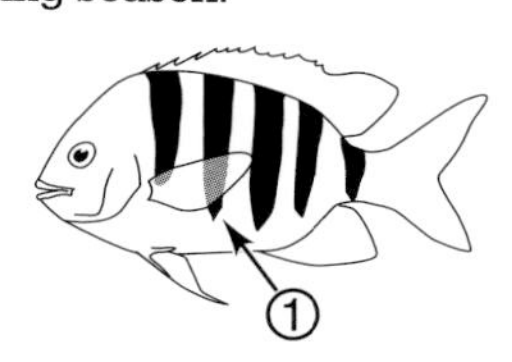

Sergeant Major Purple Phase

NIGHT SERGEANT

Abudefduf taurus

SIZE: 5 - 8 in., max. 10 in.

1. Five dark brown bars. Prefer shallow, rocky inshore surge areas.

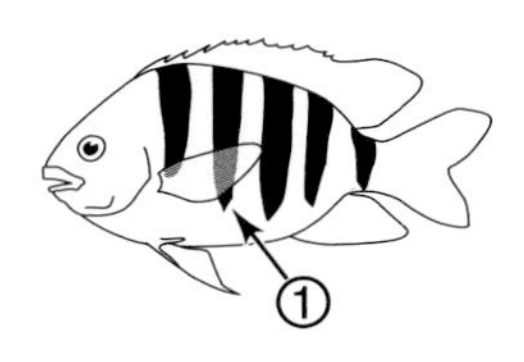

YELLOWTAIL DAMSELFISH

Microspathodon chrysurus

SIZE: 4-6 ½ in., max. 7 ½ in.

1. Yellow tail.

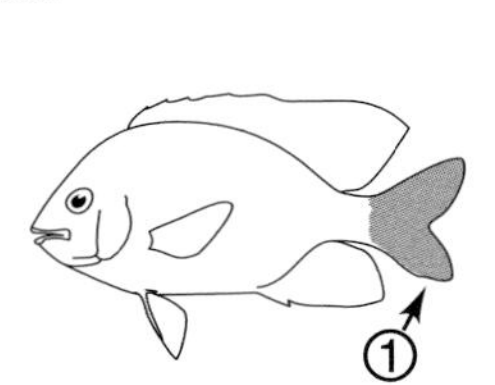

Yellowtail Damselfish Intermediate

SIZE: 2 - 4 in.

1. Brilliant blue dots on dark blue. 2. Yellow tail.

Yellowtail Damselfish Juvenile

SIZE: 1 ½ - 2 in.

1. Brilliant blue dots on dark blue. 2. Clear tail.

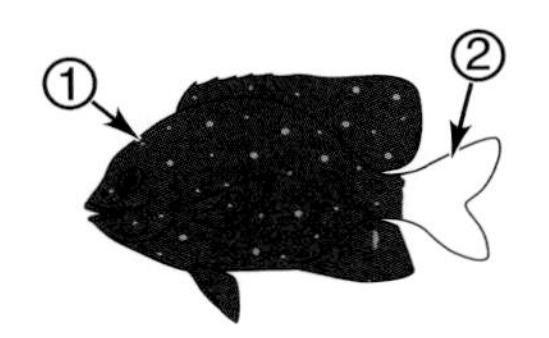

BLUE CHROMIS

Chromis cyanea

SIZE: 3-4 in., max. 5 in.

1. Slender, deeply forked tail with dark borders. Swim in midwater above reefs.

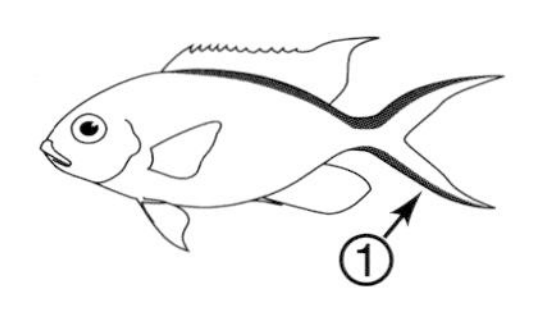

BROWN CHROMIS

Chromis multilineata

SIZE: 3-5 in., max. 6 in.

1. Border of dorsal and tail fins yellow. 2. Dark spot at base of pectoral fin.

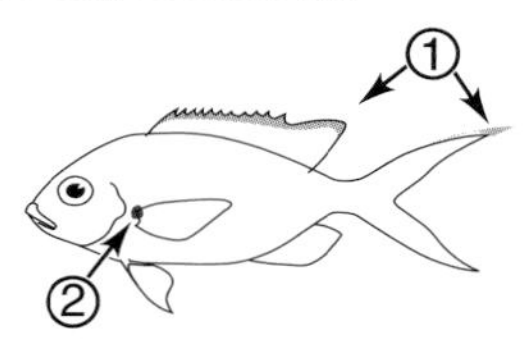

BARRED HAMLET

Hypoplectrus puella

SIZE: 3-5 in., max. 6 in.

1. Broad V-shaped bar on midbody.

BUTTER HAMLET

Hypoplectrus unicolor

SIZE: 3-4 in., max. 5 in.

1. Black saddle blotch on tail base.

FISH IDENTIFICATION GROUP 5

Heavy Body/Large Lips

Grouper/Seabass - Seabass - Soapfish - Basslet

This ID Group consists of fish with strong, well-built, "bass-like" bodies.

GROUPER/SEABASS - Serranidae - Grouper are the best known members of the seabass family. All have strong, stout bodies and large mouths. They vary in size from several feet to only one-foot. Although awkward in appearance, these fish can cover short distances quickly. Fish or crustaceans are drawn into their gullets by the powerful suction created when they open their large mouths. Grouper are hermaphroditic, beginning life as females but changing to males with maturity. Many of the larger grouper are difficult to distinguish because of their ability to radically change both colors and markings.

SEABASS - Serranidae - Often called bass, these members of the seabass family are generally more colorful than grouper. Most are small, two to four inches, and tend to be more cylindrical. All stay near the bottom. As a rule, seabass are somewhat curious and easily approached.

SOAPFISH/SEABASS - Serranidae - Soapfish are covered with mucus that produces soapsuds-like bubbles when they are caught or handled. They generally inhabit shallow waters, are solitary, reclusive night-feeders that tend to lie on the bottom and often lean against the back of a protected area.

BASSLET - Grammidae - The beautiful Fairy Basslet is the only member of this family of small fish to occasionally inhabit shallow water. They are frequently sighted near undercuts and small recesses in reefs.

NASSAU GROUPER

Epinephelus striatus

SIZE: 1-2 ft., max. 4 ft.

1. Black saddle spot on tail base. Often rest on bottom.

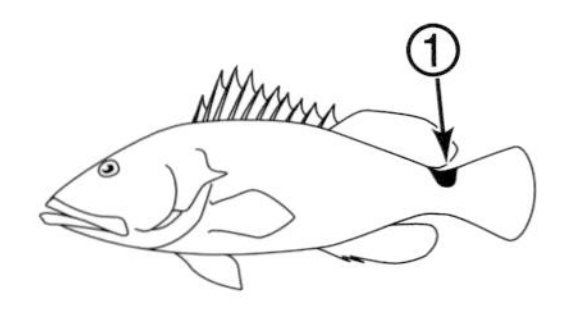

YELLOWFIN GROUPER

Mycteroperca venenosa

SIZE: 1-2 ft., max. 3 ft.

1. Outer third of pectoral fin is pale to bright yellow.

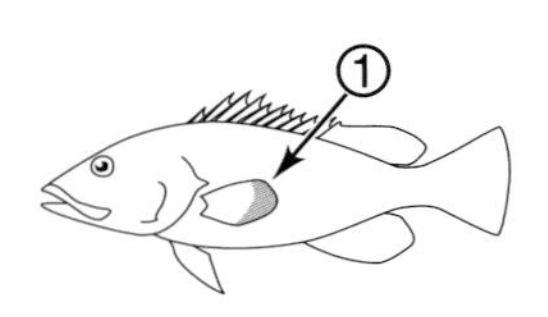

YELLOWMOUTH GROUPER

Mycteroperca interstitialis

SIZE: 1 - 2 ft., max. 2½ ft.

1. Yellow around corners of mouth. 2. Pectoral fins pale at margin.

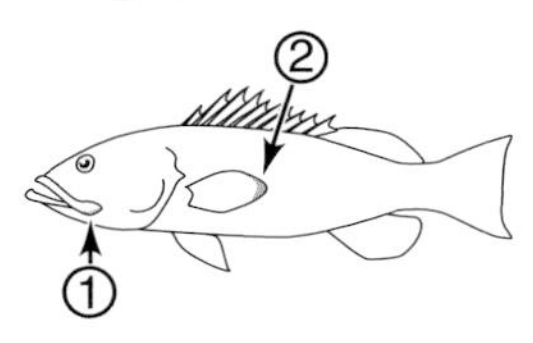

BLACK GROUPER

Mycteroperca bonaci

SIZE: 1½ - 2 ft., max. 4 ft.

1. Ends of rectangular blotches on upper body are nearly square.

TIGER GROUPER

Mycteroperca tigris

SIZE: 1 - 2 ft., max. 3½ ft.

1. Pattern of "tiger-stripe" bars. This grouper can change colors including red, and yellow.

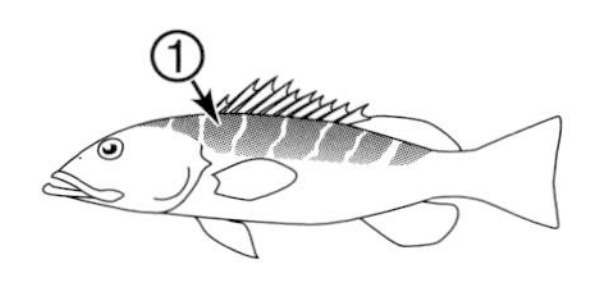

ROCK HIND

Epinephelus adscensionis

SIZE: 8 -15 in., max. 2 ft.

1. Black saddle blotch on tail base.

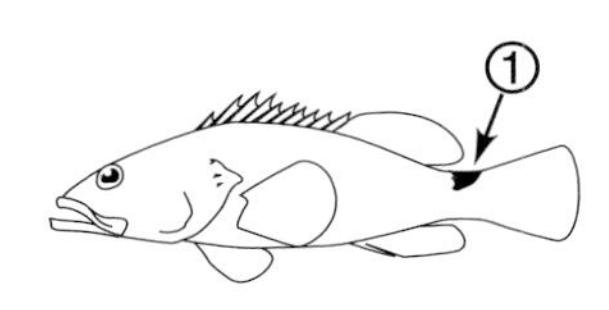

RED HIND

Epinephelus guttatus

SIZE: 10 - 14 in., max. 2 ft.

1. Tail and rear fins have black margin.

CONEY

Epinephelus fulvus

SIZE: 6 -10 in., max. 16 in.

1. Two black dots on lower lip. 2. Two black dots on upper tail base.

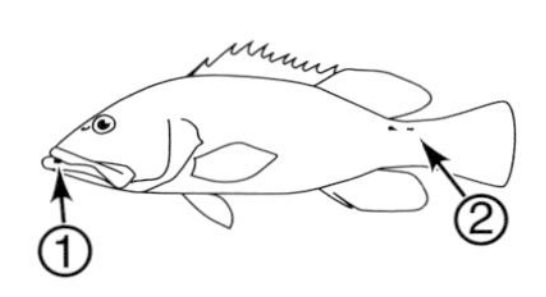

Coney Bicolor Phase

Occasionally seen in golden phase.

GRAYSBY

Epinephelus cruentatus

SIZE: 6 - 10 in., max. 1 ft.

1. Three to five pale or dark dots along base of dorsal fin. 2. Tail rounded.

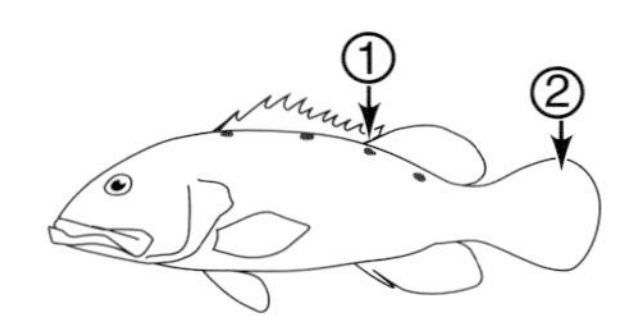

HARLEQUIN BASS

Serranus tigrinus

SIZE: 2 1/2 - 3 1/2 in., max. 4 in.

1. Series of dark bars on body.

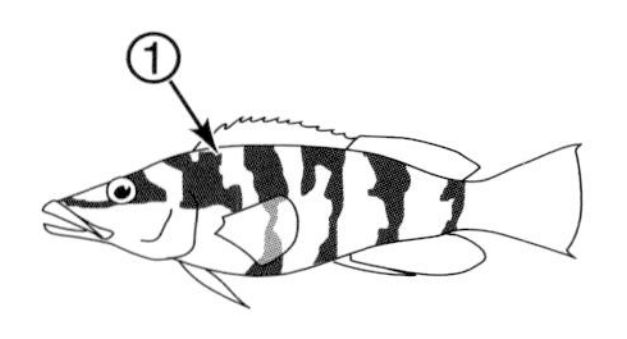

SAND PERCH

Diplectrum formosum

SIZE: 5 - 9 in., max. 1 ft.

1. Pale blue lines on head. 2. Thin, blue body stripes.

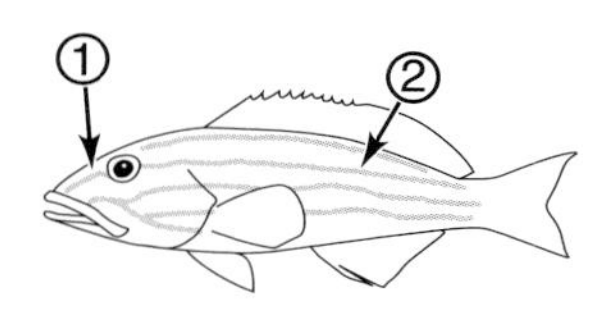

GREATER SOAPFISH

Rypticus saponaceus

SIZE: 5 - 9 in., max. 13 in.

Solitary and inactive; often rest on bottom under ledges.

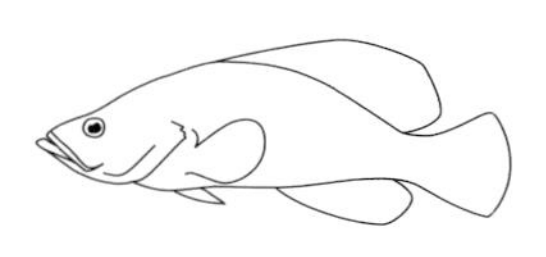

FAIRY BASSLET

Gramma loreto

SIZE: 1 1/2 - 2 1/2 in., max. 3 in.

1. Bicolored. 2. Dark spot on foredorsal fin.

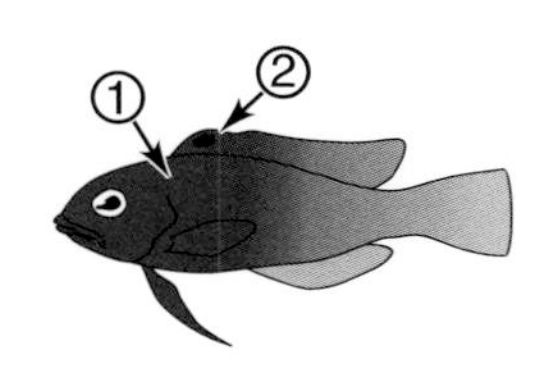

FISH IDENTIFICATION GROUP 6

Swim with Pectoral Fins/Obvious Scales

Parrotfish-Wrasse-Hogfish/Wrasse-Razorfish/Wrasse

This ID Group consists of fish that primarily use their pectoral fins to swim. They have even rows of large, noticeable scales and beak-like mouths.

PARROTFISH - Scaridae - Powerful jaws, fused teeth or "beaks", and bright colors give parrotfish their common name. They are among the most common large fish seen on reefs. Most species are solitary. Their "beaks" are used to scrape algae and polyps from coral and rocks. In the process, large amounts of coral (limestone) are taken in and ground in the gullets to extract bits of algae and occasionally polyps. Clouds of the chalky residue are regularly excreted as the fish move about the reef.

Identification of parrotfish is made difficult due to the dramatic changes in shape, color and markings that occur in most species as they mature. The phases include the JUVENILE, INITIAL, and TERMINAL PHASE, which is the largest and most colorful. The initial phase includes sexually mature females and, in some species, immature males. Only sexually mature males are in the terminal phase. The first photograph of each parrotfish species shows the terminal phase.

WRASSE - Labridae - Wrasse are prolific shallow reef inhabitants. They are closely related and similar to parrotfish, but are generally much smaller and have more elongated "cigar" shapes. Their noticeable front teeth are used to obtain food by crushing the shells of invertebrates, such as sea urchins. By day, wrasse swim busily in loose, often mixed aggregations, over coral reefs and sand. A few species inhabit grass beds. Like parrotfish, wrasse go through similar changes in color, shape and markings during maturation.

HOGFISH, RAZORFISH/WRASSE - Labridae - Hogfish and razorfish are members of the wrasse family, but because of their unique shapes have acquired separate common names. Hogfish have long snouts which they use to root for food.

Razorfish are named for their blunt, highly compressed "razor-like" heads that are shaped something like the prow of a ship. When frightened, the small three to five inch fish dive into the sand and tunnel away.

MIDNIGHT PARROTFISH

Scarus coelestinus

SIZE: 1 - 2 ft., max. 3 ft.

1. Bright blue markings on head.

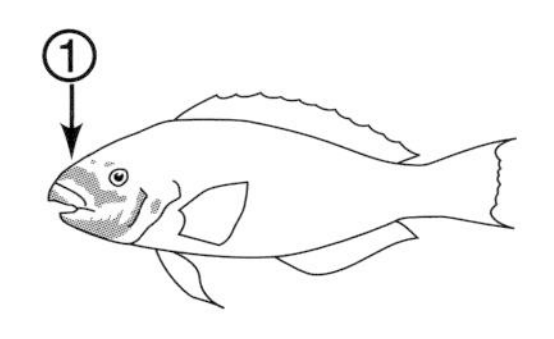

RAINBOW PARROTFISH

Scarus guacamaia

SIZE: 1 1/2 - 3 ft., max. 4 ft.

1. Orange-brown head. 2. Bright green rear body.

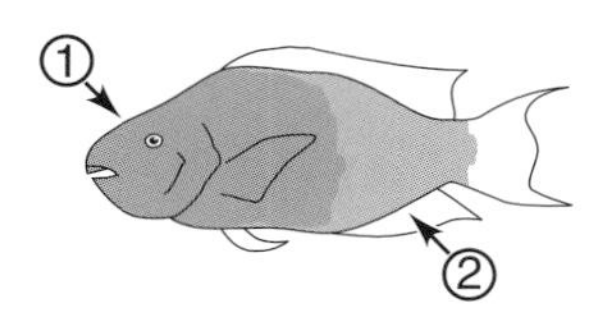

BLUE PARROTFISH

Scarus coeruleus

SIZE: 1 - 3 ft., max. 4 ft.

Powder blue overall.

QUEEN PARROTFISH

Scarus vetula

SIZE: 12 - 16 in., max. 2 ft.

1. Dramatic blue to green markings around mouth.

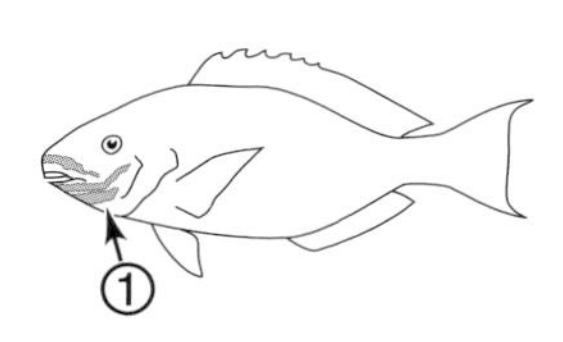

Queen Parrotfish Initial Phase

SIZE: 2 1/2 - 12 in.

1. Broad, white midbody stripe .

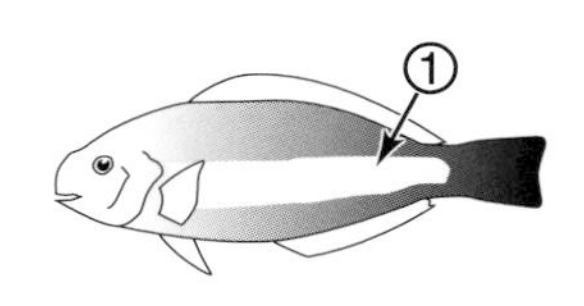

STOPLIGHT PARROTFISH

Sparisoma viride

SIZE: 1 - $1\frac{1}{2}$ ft., max. 2 ft.

1. Yellow spot at upper corner of gill cover.
2. Crescent on tail.

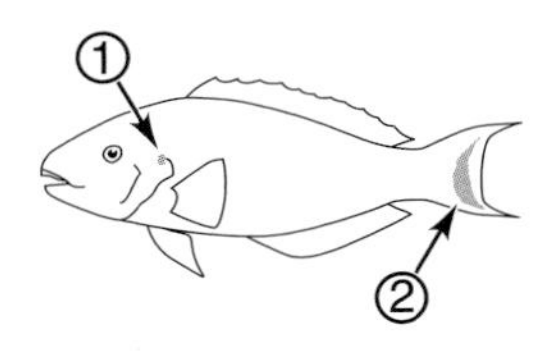

Stoplight Parrotfish Initial Phase

SIZE: 5 - 10 in., max. 1 ft.

1. Red belly and tail.

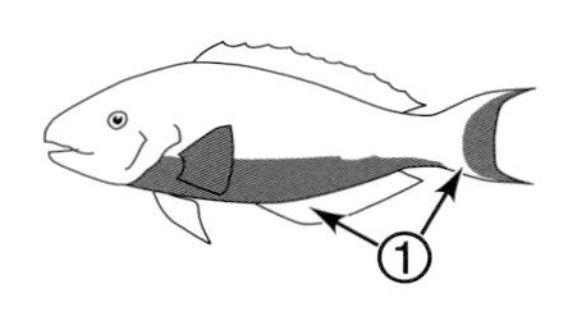

Stoplight Parrotfish Juvenile

SIZE: 2 - 4 in., max. 5 in.

1. Three rows of spots run length of body.

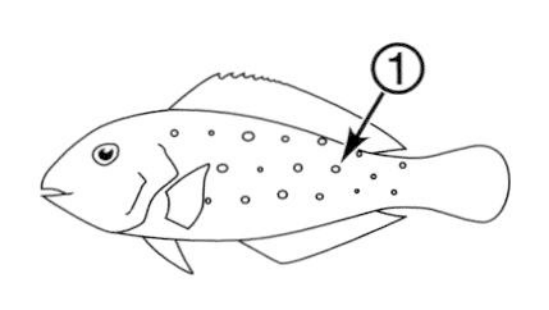

REDBAND PARROTFISH

Sparisoma aurofrenatum

SIZE: 6 - 10 in., max. 11 in.

1. Yellow blotch with black dots on upper forebody. 2 Colored line below eye.

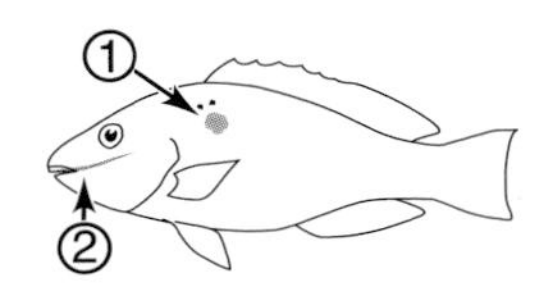

Redband Parrotfish
Initial Phase

Solid body color with red fins.

1. White spot behind dorsal fin.

SIZE: 2 1/2 - 5 in., max. 6 in.

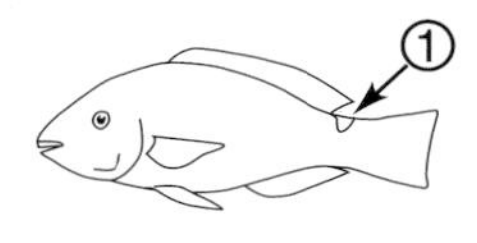

Redband Parrotfish
Juvenile

SIZE: 1 1/2 - 2 1/2 in., max. 3 in.

1. Two white stripes. 2. Large black blotch behind gill cover.

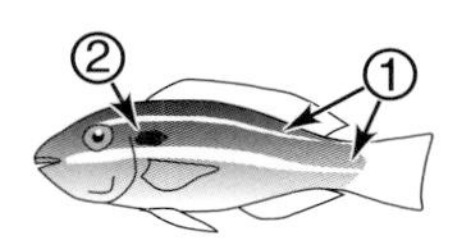

PRINCESS PARROTFISH

Scarus taeniopterus

SIZE: 8 - 10 in., max. 13 in.

1. Borders on top and bottom margins of tail.

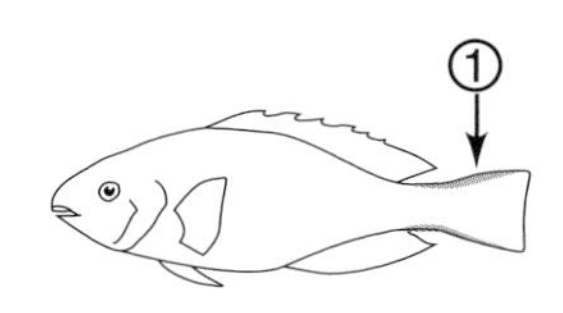

Princess Parrotfish
Juvenile

SIZE: 2 - 4 in.

1. Borders of tail dark.

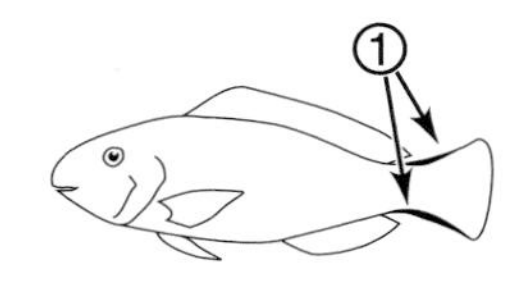

STRIPED PARROTFISH

Scarus croicensis

SIZE: 8 - 9 in., max. 10 in.

1. Series of thin stripes on tail.

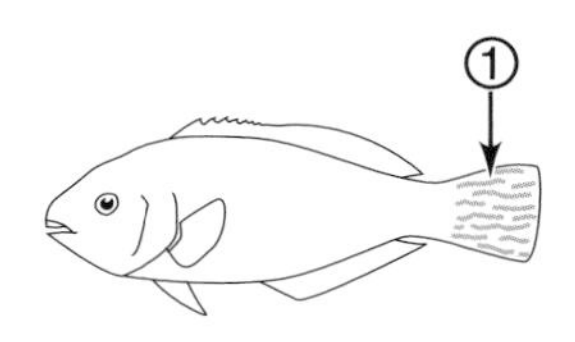

Striped Parrotfish Juvenile

SIZE: 2 - 4 in.

1. No borders or markings on tail.

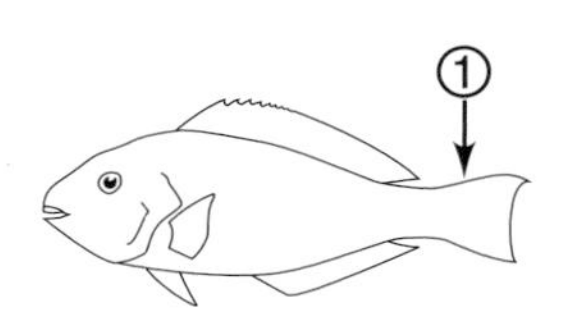

REDTAIL PARROTFISH

Sparisoma chrysopterum

SIZE: 14 - 16 in., max. 18 in.

1. Blue wash behind pectoral fin. 2. Reddish crescent on tail margin.

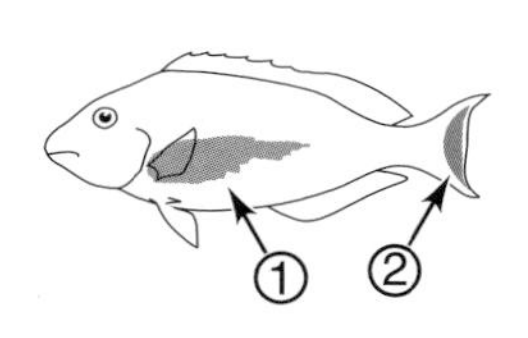

Redtail Parrotfish Initial Phase

SIZE: 8 - 14 in., max. 16 in.

Reddish gray. Can rapidly fade, intensify or change color.

YELLOWTAIL PARROTFISH

Sparisoma rubripinne

SIZE: 8 - 14 in., max. 1 1/2 ft.

1. Center of tail yellowish

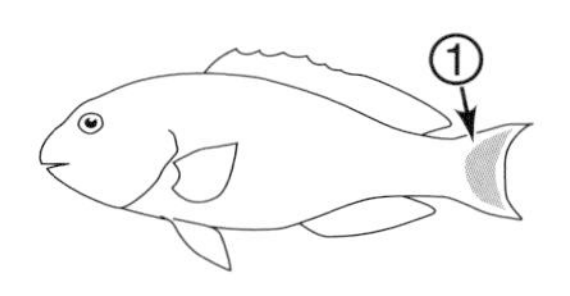

Yellowtail Parrotfish Initial Phase

SIZE: 8 - 12 in., max. 14 in.

1. Yellow tail.

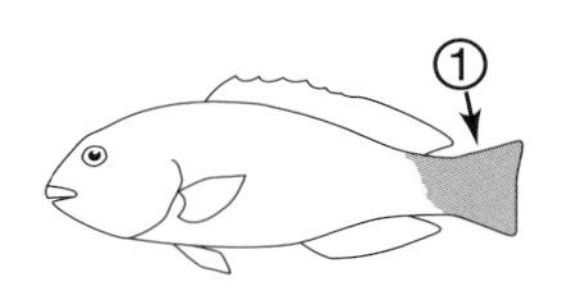

PUDDINGWIFE

Halichoeres radiatus

SIZE: 12 - 15 in., max 18 in.

1. Dark dot at base of pectoral fin.

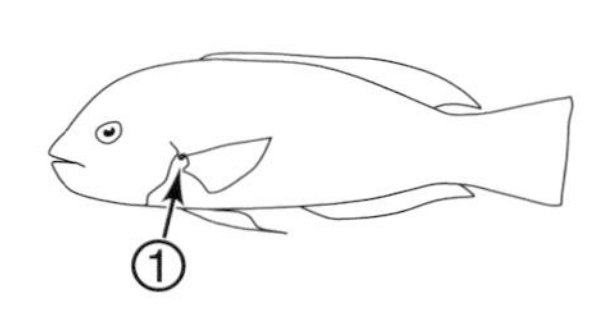

Puddingwife Juvenile

SIZE: 1 1/2 - 2 1/2 in.

1. Large black spot ringed in blue on midback and dorsal fin.

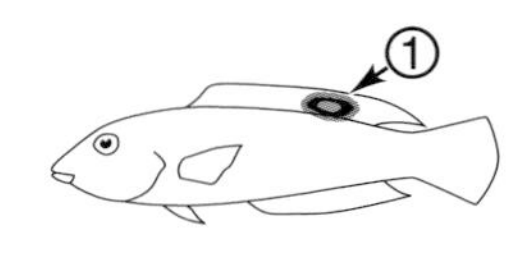

HOGFISH

Lachnolaimus maximus

SIZE: 1 - 2 ft., max. 3 ft.

1. First three spines of dorsal fin are long. Vary from white to mottled or banded reddish brown.

SPANISH HOGFISH

Bodianus rufus

SIZE: 8 -14 in., max. 2 ft.

1. Purple upper forebody. 2. Yellow belly and tail.

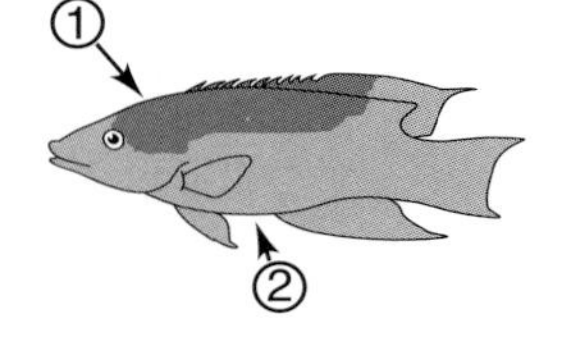

BLUEHEAD

Thalassoma bifasciatum

SIZE: 4- 5 in., max. 6 in.

1. Blue head. Very common species.

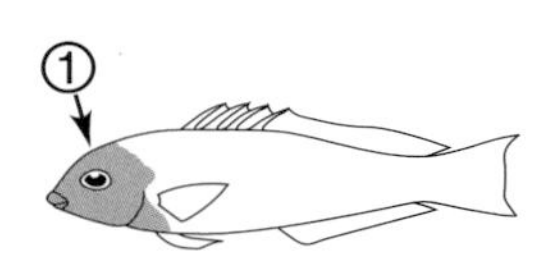

Bluehead Initial Phase

SIZE: 3 - 5 in.

1. Dark blotch on foredorsal fin.

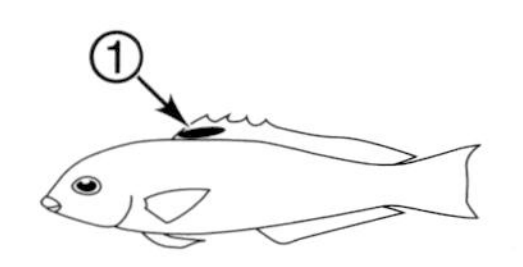

Bluehead Juvenile

SIZE: 1 - 4 in.

1. Dark blotch on foredorsal fin

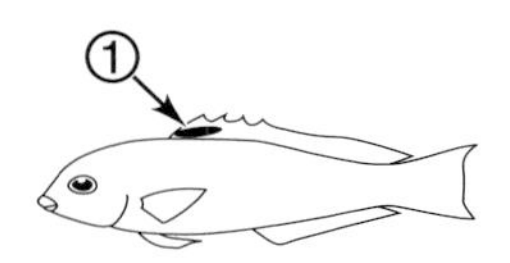

YELLOWHEAD WRASSE

Halichoeres garnoti

SIZE: 5 - 6 in., max. 8 in.

1. Yellow head and forebody. Very common species.

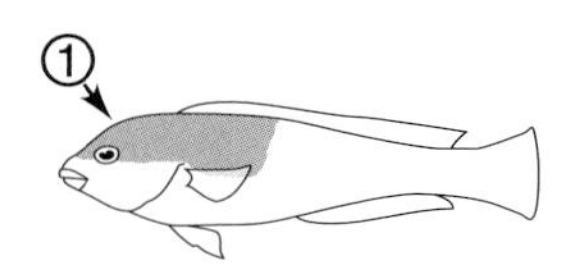

Yellowhead Wrasse Initial Phase

SIZE: 3 - 4 in.

1. Yellow midbody wash.

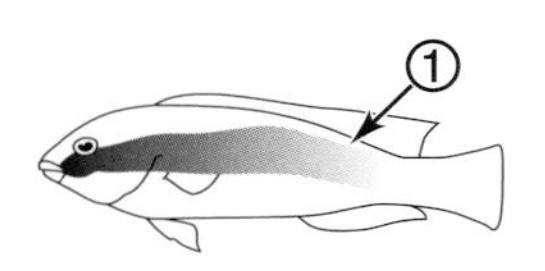

Yellowhead Wrasse Juvenile

SIZE: 1 1/2 - 2 1/2 in.

1. Brilliant blue midbody stripe.

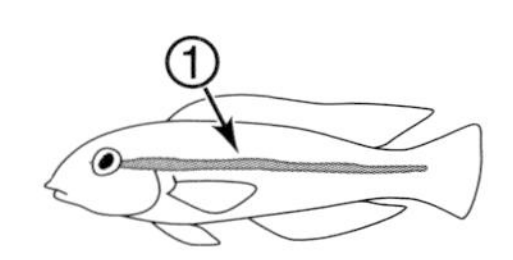

CLOWN WRASSE

Halichoeres maculipinna

SIZE: 3 - 5 in., max. 6 1/2 in.

1. Red lines across top of head. 2. Dark blotch on side (occasionally absent).

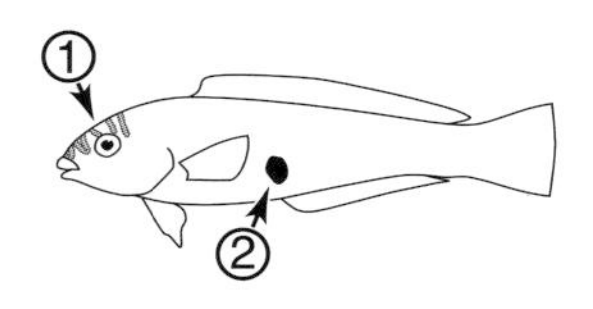

Clown Wrasse Juvenile

SIZE: 2 - 3 in.

1. Yellow stripe with broad black stripe below.

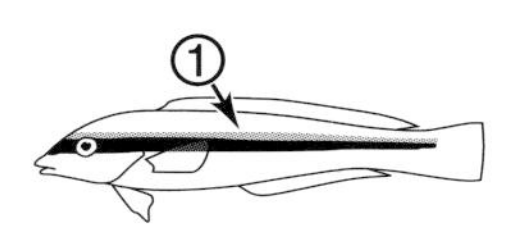

SLIPPERY DICK

Halichoeres bivittatus

SIZE: 5 - 7 in., max. 9 in.

1. Small green and yellow spot above pectoral fin.

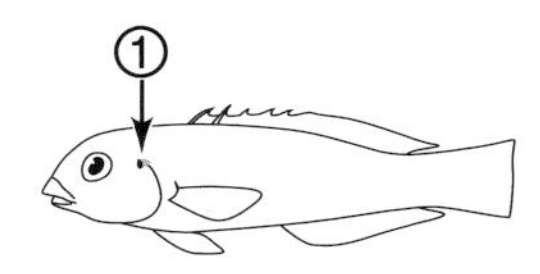

Slippery Dick Juvenile

SIZE: 1 1/2 - 3 in.

1. Small green and yellow spot above pectoral fin. White when over sand. Brownish when over reef.

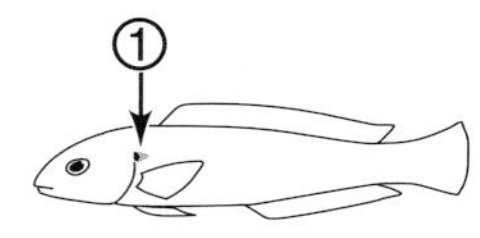

GREEN RAZORFISH

Hemipteronotus splendens

SIZE: 2 1/2 - 4 in., max. 5 1/2 in.

1. Dark spot (occasionally two) at midbody. Thin body. Hover above sand. Dive into sand when threatened.

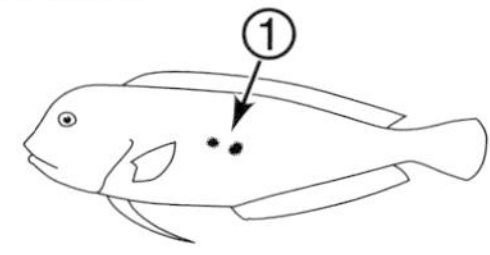

Green Razorfish Female

SIZE: 1 1/2 - 3 1/2 in.

Color and markings variable. Often display bars. Hover over sand with body curved in S-shape.

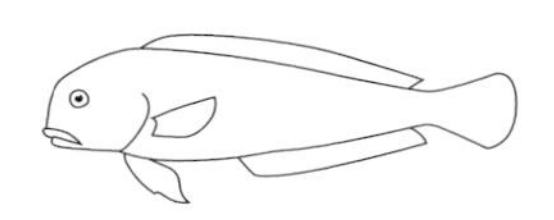

ROSY RAZORFISH

Hemipteronotus martinicensis

SIZE: 3 - 4 1/2 in., max. 6 in.

1. Dark spot at base of pectoral fin. Thin body. Hover above sand. Dive into sand when threatened.

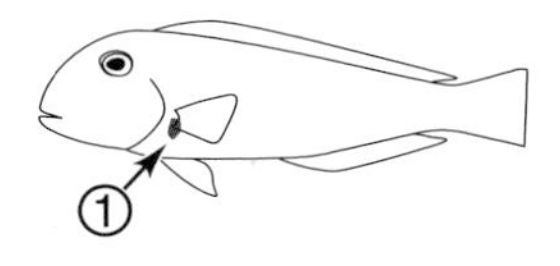

Rosy Razorfish Female

SIZE: 2 1/2 in., max. 4 1/2 in.

1. White belly patch. 2. Darkish bar on gill cover.

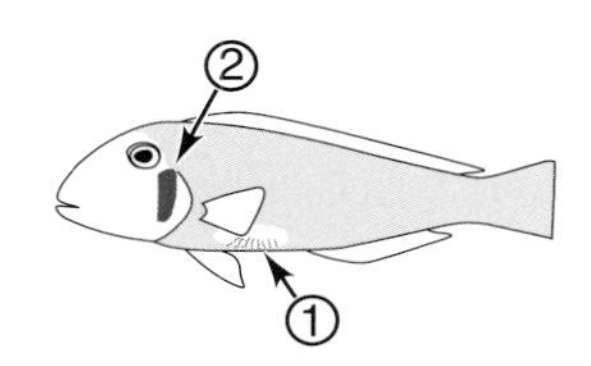

FISH IDENTIFICATION GROUP 7

Reddish/Big Eyes

Squirrelfish - Bigeye - Cardinalfish

This ID Group consists of reddish fish with large eyes.

SQUIRRELFISH - Holocentridae - This family of reddish fish, with large "squirrel-like" eyes and long, pronounced rear dorsal fins, are common inhabitants of shallow reefs. Although nocturnal, most are spotted in the openings of reef pockets during the day.
BIGEYE - Priacanthidae - The Glasseye Snapper, which is not a snapper at all, is the only member of the bigeye family that inhabits shallow waters.

SQUIRRELFISH

Holocentrus adscensionis

SIZE: 6 - 12 in., max. 16 in.

1. Yellowish foredorsal fin.

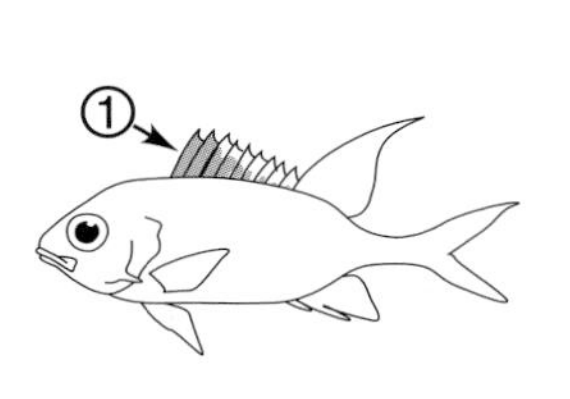

LONGSPINE SQUIRRELFISH

Holocentrus rufus

SIZE: 5 - 10 in., max. 12 in.

1. White markings at tips of dorsal fin spines.

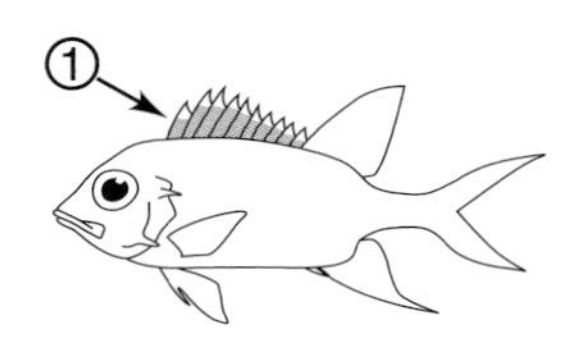

REEF SQUIRRELFISH

Holocentrus coruscum

SIZE: 3 1/2 - 5 in., max. 6 in.

1. Black blotch on foredorsal fin.

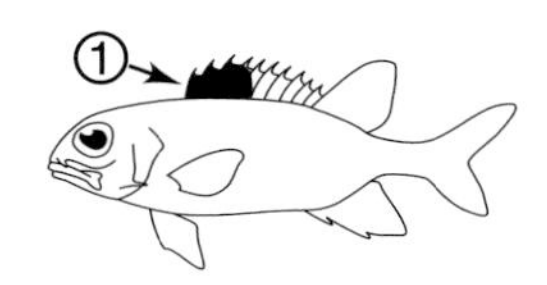

DUSKY SQUIRRELFISH

Holocentrus vexillarius

SIZE: 3 - 5 in., max. 7 in.

1. Anal and tail fins bordered in brownish red.

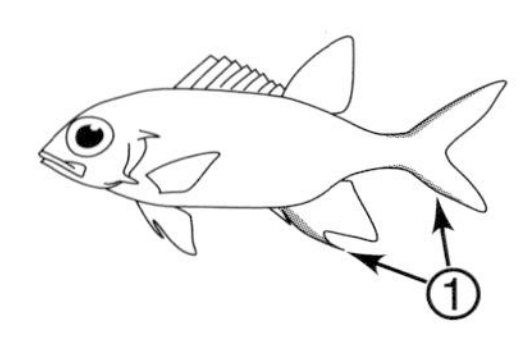

BLACKBAR SOLDIERFISH

Myripristis jacobus

SIZE: 3½ - 5½ in., max. 8½ in.

1. Black bar behind head.

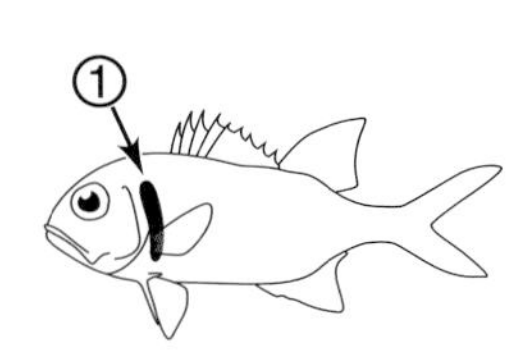

GLASSEYE SNAPPER

Priacanthus cruentatus

SIZE: 7 - 10 in., max. 1 ft.

1. Silver bars on back, which may be faint.

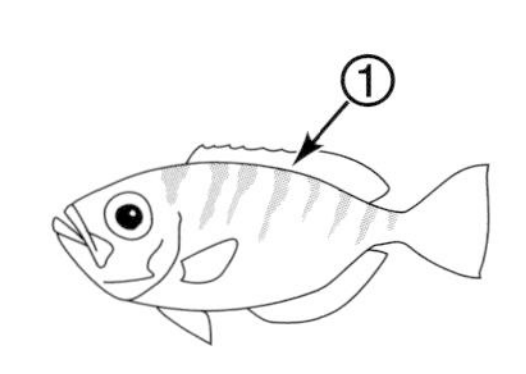

BELTED CARDINALFISH

Apogon townsendi

SIZE: 1¼ - 2 in., max. 2½ in.

1. Dark bar from rear dorsal fin to anal fin and two bars at base of tail.

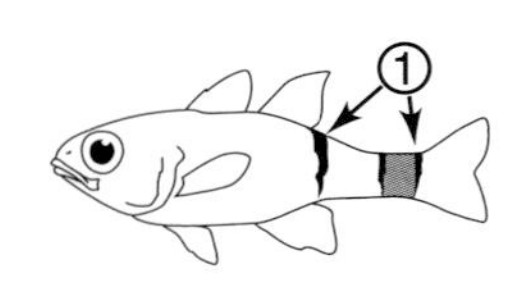

FISH IDENTIFICATION GROUP 8

Bottom-Dwellers

Goby - Blenny - Jawfish - Flounder - Scorpionfish - Others

This ID Group consists of fish that live on the surface of coral, rock, sponge or sand.

GOBY - Gobiidae - Gobies are the smallest members of this ID Group (generally only one to two inches long). Most rest on their pectoral and ventral fins. Gobies and blennies are often confused, but can easily be distinguished by their dorsal fins — gobies have two, while most blennies have one. Another observable difference is the tendency of gobies to rest in a stiff, straight position, while blennies are more flexed and curved.

BLENNY - Blenniidae & Clinidae - Blennies are also small bottom-dwellers. Most blennies have fleshy appendages, called cirri, above their eyes. Because of their small size and ability to change colors and markings to blend with the background, they often go unnoticed. Several species live in holes and are only seen with their heads exposed.

JAWFISH - Opistognathidae - By far, the most common species of jawfish sighted by snorkelers is the Yellowhead. These 2 to 4 inch fish hover vertically in the water column just above their burrows in the sand. They live in groups from a few to several dozen. When closely approached, they enter the burrows, tail first. The males have the curious trait of incubating eggs in their mouths.

LEFTEYE FLOUNDER - Bothidae - Flounders are unique, flat fish that actually lie on their sides, not their stomachs. Within a few weeks of birth, the eye on the bottom slowly migrates to the exposed side. Their exposed pectoral fin is more like a dorsal fin, while the dorsal and anal fins almost ring the rounded body.

LIZARDFISH - Synodontidae - Sand Divers are the most common member of the lizardfish family. All have large, upturned mouths, pointed snouts and long, cylindrical bodies. Experts at camouflage, they rest motionless on the bottom, blending with their surroundings. Some bury themselves in the sand with only their heads protruding as they wait for unsuspecting prey.

FROGFISH - Antennariidae - Frogfish are globular with large, extremely upturned mouths which can be opened to the width of their bodies to engulf prey. The first dorsal spine, located on the snout, has evolved into a thin, stalk-like structure which is used as a lure to attract small fish. Masters at camouflage, they can change to virtually any color to match the background. Because they seldom move they appear to be sponges or clumps of algae.

SEAHORSE - Syngnathidae - These strange little fish have trumpet-like snouts and small mouths. Their bodies are encased in protective bony rings which are quite apparent. They have cocked heads and hold themselves upright by coiling their elongated tails around a hold-fast such as a gorgonian.

SCORPIONFISH - Scorpaenidae - Fleshy appendages, or flaps, help camouflage scorpionfishes' large heads and stocky bodies. Mottled and spotted in earthtones, they are difficult to detect as they lie motionless on bottom rubble or patches of algae. Like most ambush predators, they have large mouths that open quickly to suck in unsuspecting prey. Spines of the foredorsal fin, which can be raised defensively, are venomous and can inflict a painful wound; however, the fish are not aggressive and mishaps with snorkelers or divers are rare and seldom, if ever, fatal.

BRIDLED GOBY

Coryphopterus glaucofraenum

SIZE: 1 1/2 - 2 1/2 in., max. 3 in.

1. Short stripe from mouth to gill cover.
2. Series of dots along base of dorsal fin.

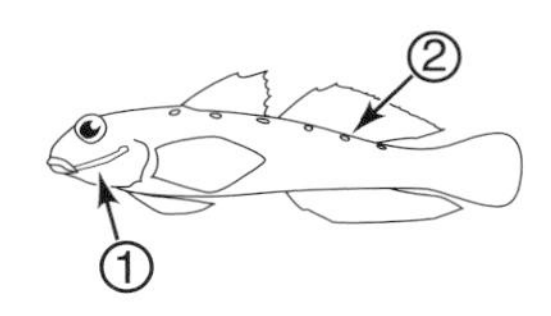

GOLDSPOT GOBY

Gnatholepis thompsoni

SIZE: 1 1/2 - 2 1/2 in., max. 3 in.

1. Dark "mask" runs across head and eyes.
2. Gold spot above pectoral fin.

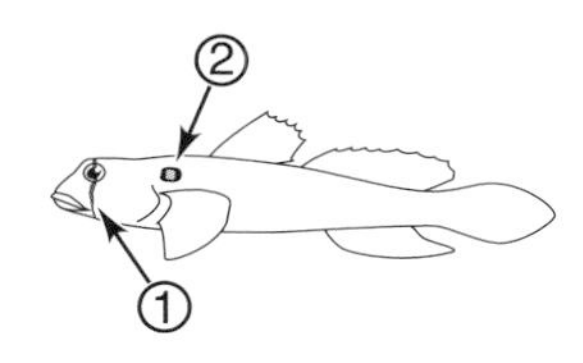

COLON GOBY

Coryphopterus dicrus

SIZE: 1 1/2 - 2 in., max. 2 1/2 in.

1. Two darkish spots at base of pectoral fin.

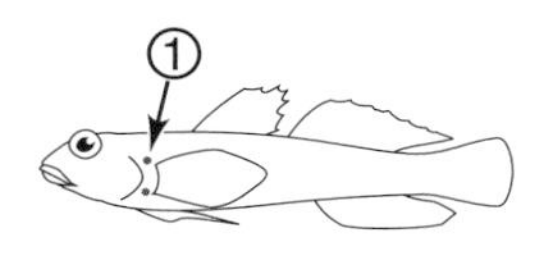

PALLID GOBY

Coryphopterus eidolon

SIZE: 1 - 2 in., max. 2 in.

1. Yellowish stripe extends from rear of eye.

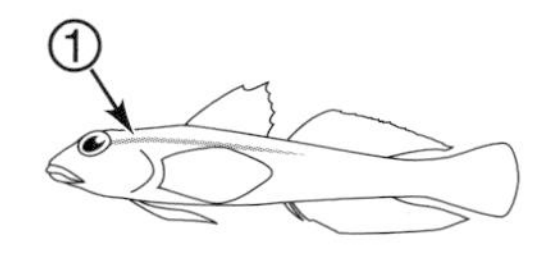

NEON GOBY

Gobiosoma oceanops

SIZE: 1-1½ in., max. 2 in.

1. Electric blue body stripe runs from front of eye to tail base. Perch on coral heads.

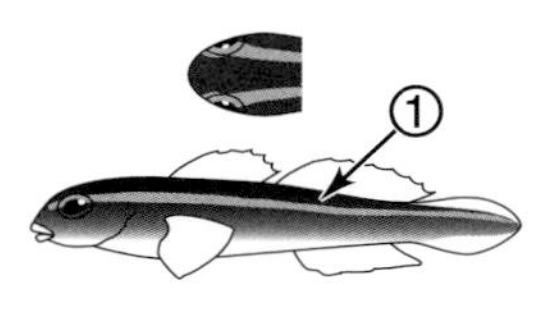

CLEANING GOBY

Gobiosoma genie

SIZE: 1-1½ in., max. 2 in.

1. Yellow "V" on snout fades into pale body stripes. Perch on coral heads.

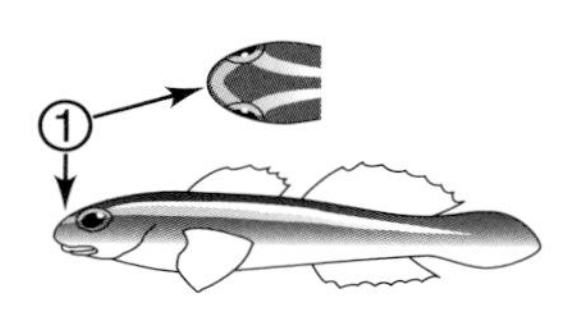

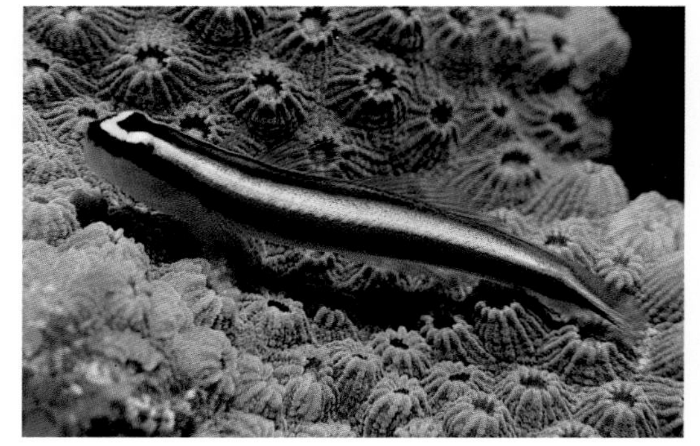

SEAWEED BLENNY

Parablennius marmoreus

SIZE: 1½ - 3 in., max 4 in.

1. Blue vertical line markings on face. Markings and colors vary greatly.

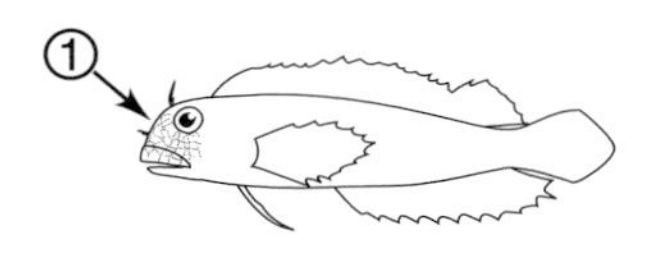

SADDLED BLENNY

Malacoctenus triangulatus

SIZE: 1½ - 2 in., max 2½ in.

1. Four dark, inverted, triangluar markings across back.

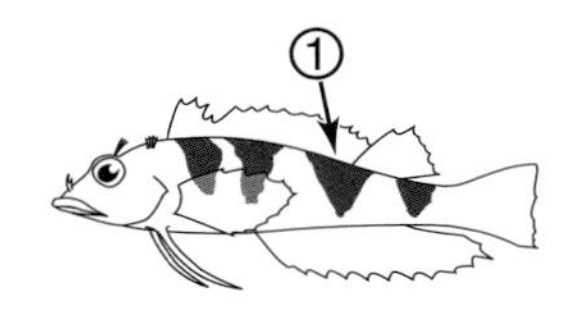

ROSY BLENNY

Malacoctenus macropus

SIZE: 1 1/2 - 2 in., max 2 1/2 in.

1. Red spots and markings on underside of head and cheeks.

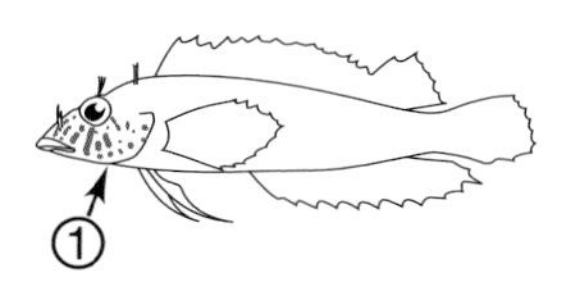

Rosy Blenny Female

SIZE: 1 1/2 - 2 in., max 2 1/2 in.

No red spots or dependable markings. Can have dark bars on back.

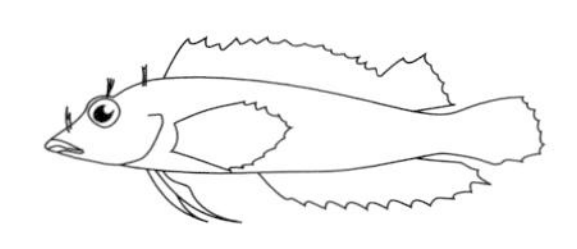

SAILFIN BLENNY

Emblemaria pandionis

SIZE: 1 1/2 - 2 in., max 2 1/2 in.

1. Unusually high dorsal fin. 2. Series of blue dots on face. Photo: Male displaying to attract female.

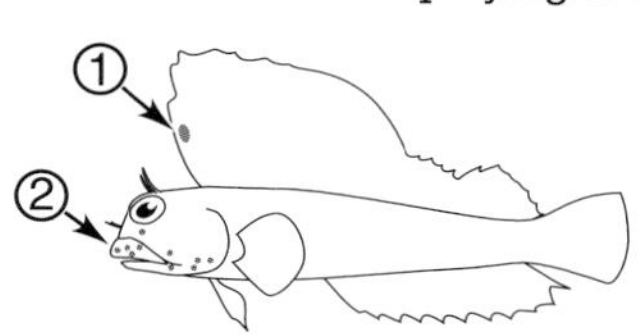

Sailfin Blenny Female

SIZE: 1 1/2 - 2 in., max 2 1/2 in.

1. Thin, broken diagonal lines on foredorsal fin.

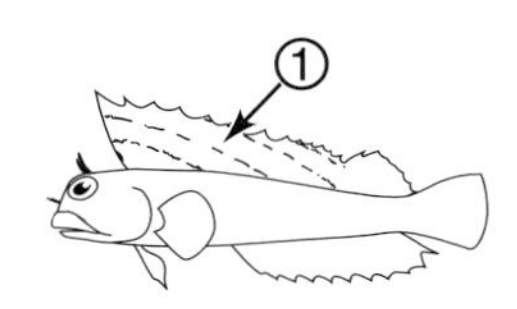

REDLIP BLENNY

Ophioblennius atlanticus

SIZE: 2½-4½ in., max. 5 in.

1. Large reddish lips. 2. Blunt head.

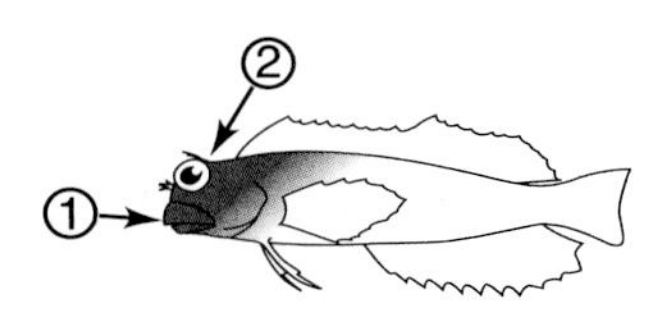

YELLOWHEAD JAWFISH

Opistognathus aurifrons

SIZE: 2-3 in., max. 4 in.

1. Yellowish head. Hover in vertical position above burrows. Retreat into hole when approached.

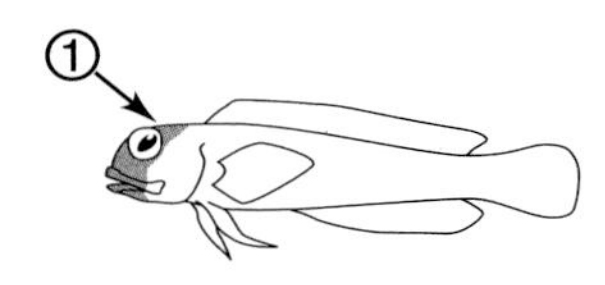

PEACOCK FLOUNDER

Bothus lunatus

SIZE: 6-15 in., max. 18 in.

1. Unusually long pectoral fin, often erect. 2. Blue spots on fins and head.

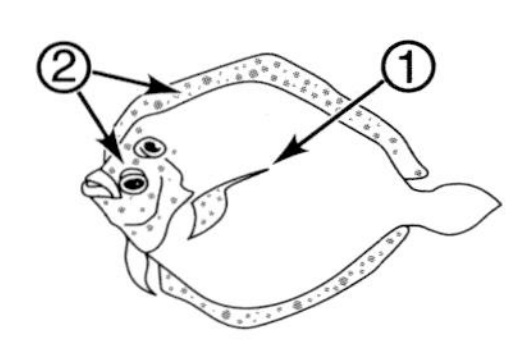

SAND DIVER

Synodus intermedius

SIZE: 4-14 in., max. 18 in.

1. Dark spot at upper end of gill cover. 2. Thin yellow body stripes.

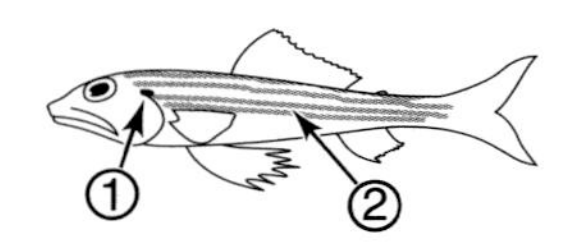

FLYING GURNARD

Dactylopterus volitans

SIZE: 6 - 14 in., max. 18 in.

1. Huge, fan-like pectoral fins look like wings.

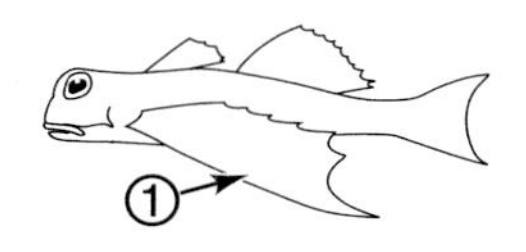

SPOTTED SCORPIONFISH

Scorpaena plumieri

SIZE: 7 - 14 in., max. 18 in.

1. Three dark bars on tail. Seldom move; blend with background.

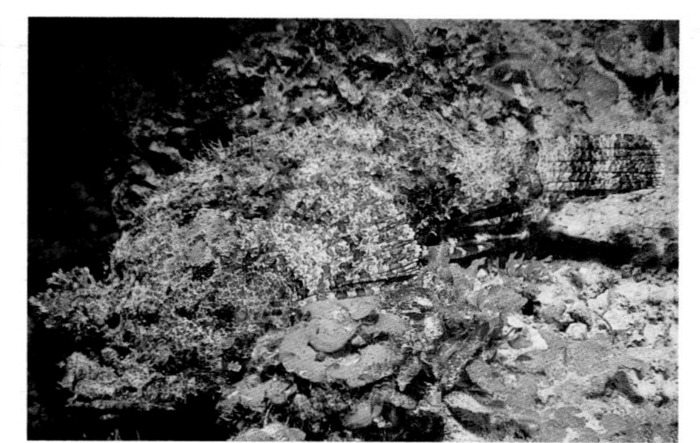

LONGLURE FROGFISH

Antennarius multiocellatus

SIZE: 3 - 5 in., max. 8 in.

1. Three spots on tail. Many color phases. Difficult to sight because they seldom move and camouflage well.

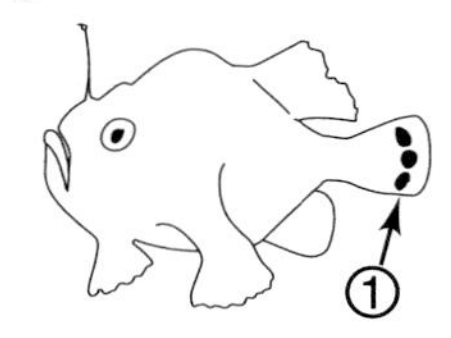

LONGSNOUT SEAHORSE

Hippocampus reidi

SIZE: $2^1/_2$ - 4 in., max. 6 in.

1. Black specks over head and body. Curl tail around gorgonians, occasionally float free.

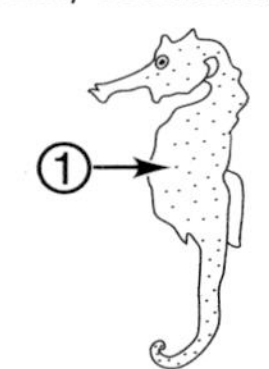

FISH IDENTIFICATION GROUP 9

Odd-Shaped Swimmers

Puffer-Boxfish-Triggerfish & Filefish - Drum - Others

This ID Group consists of swimming fish that do not have a typical fish-like shape.

TRUMPETFISH - Aulostomidae - These common predators can change colors, including brown, blue-gray, or even yellow, and positions to blend with the background. Often float head down, paralleling stalks of sea rods. Occasionally attempt to camouflage themselves by hovering just above larger fish.

GOATFISH - Mullidae - Both species of goatfish in the region are common. They use barbels to dig in sand and around areas of rubble for food. The Spotted Goatfish rapidly and dramatically changes color from white to blotched and mottled reddish brown when resting on the bottom. Yellow Goatfish often form large schools. When not feeding often mix with grunts and snappers in the protection of the reef.

SMOOTH & SPINY PUFFER - Tetraodontidae - Puffers have the unique ability to draw in water to greatly inflate their bodies as a defense. The family is divided into two basic groups: smooth puffers, with smooth, spineless skin; and spiny puffers, with a covering of stout spines.

BOXFISH - Ostraciidae - Boxfish are protected by a triangular, bony box of armor. They have small protrusible mouths and broom-like tails. They are divided into two groups: cowfish, which have a sharp spine over each eye; and trunkfish, which do not have these spines. These relatively slow swimmers move with a sculling action of their dorsal, anal and pectoral fins. The tail fin is only brought into play when greater speed is desired.

TRIGGERFISH & FILEFISH - Balistidae - Each family member has an elongated front dorsal spine that can be raised and lowered. The family is divided into two groups: triggerfish, that can lock their stout front dorsal fin into place; and filefish, that cannot lock their elongated spine.

DRUM - Sciaenidae - The common name of these fish is derived from their ability to vibrate their swim bladders, thereby producing a low-pitched, resonant sound. Reef-dwelling drums, especially juveniles, have unusually elongated foredorsal fins, making them quite distinctive. All are similarly patterned in white and black, but are easily distinguished with attention to detail.

TRUMPETFISH

Aulostomus maculatus

SIZE: 1½-2½ ft., max. 3 ft.

1. Trumpet-like mouth. Change colors.

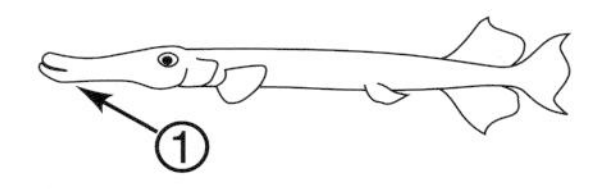

SAND TILEFISH

Malacanthus plumieri

SIZE: 1 - 1½ ft., max. 2 ft.

1. Dark area on upper central tail. Hover near entrance of burrows in sandy areas.

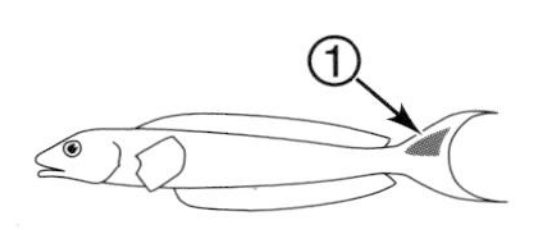

GLASSY SWEEPER

Pempheris schomburgki

SIZE: 3 - 5 in., max. 6 in.

1. Dark band at base of anal fin. Drift in caves often in large groups.

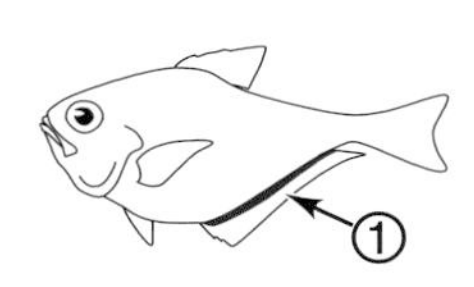

YELLOW GOATFISH

Mulloidichthys martinicus

SIZE: 6 - 12 in., max. 16 in.

1. Yellow tail and midbody stripe. Often in large schools.

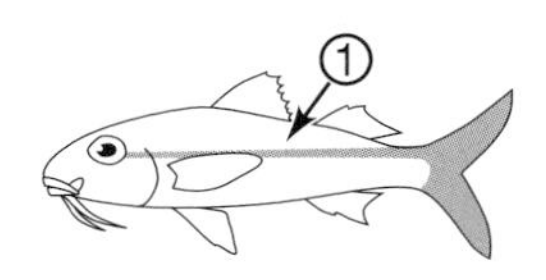

SPOTTED GOATFISH

Pseudupeneus maculatus

SIZE: 5 - 8 in., max. 11 in.

1. Three dark, rectangular body blotches.

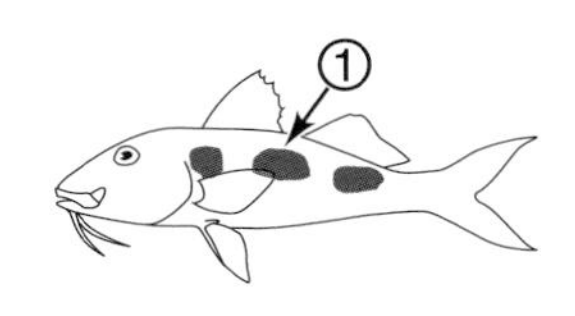

SHARPNOSE PUFFER

Canthigaster rostrata

SIZE: 2 - 3½ in., max. 4½ in.

1. Blue lines radiate from around eye. Swim close to protection of reef.

BANDTAIL PUFFER

Sphoeroides spengleri

SIZE: 4 - 7 in., max. 1 ft.

1. Two dark bands on tail. 2. Row of dark blotches along abdomen.

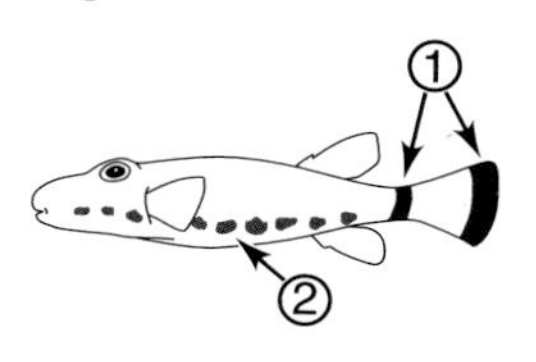

PORCUPINEFISH

Diodon hystrix

SIZE: 1 - 2 ft., max. 3 ft.

1. Spots on fins.

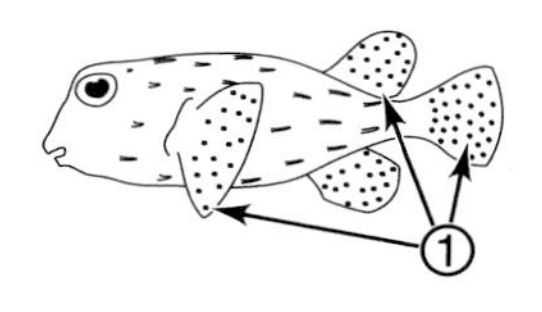

BALLOONFISH

Diodon holocanthus

SIZE: 8 - 14 in., max. 20 in.

1. Long spines on head.

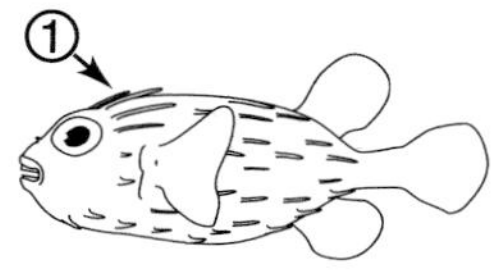

HONEYCOMB COWFISH

Lactophrys polygonia

SIZE: 7 - 15 in., max. 18 in.

1. Honeycomb pattern on body.

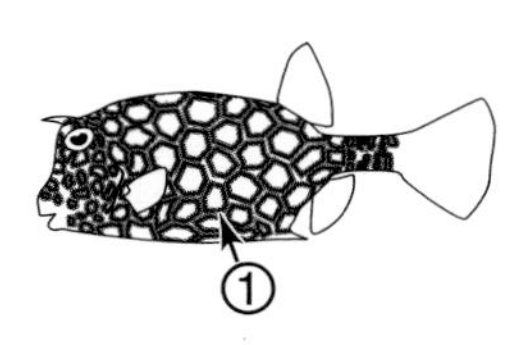

SCRAWLED COWFISH

Lactophrys quadricornis

SIZE: 8 - 15 in., max. 18 in.

1. Scrawled pattern of bluish markings covers body.

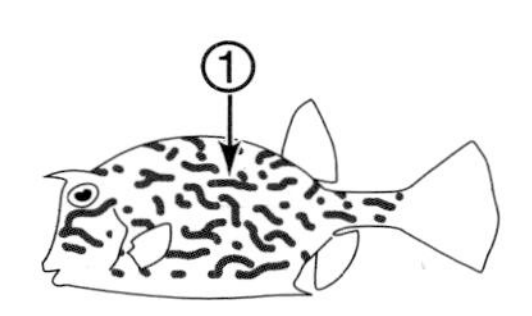

SMOOTH TRUNKFISH

Lactophrys triqueter

SIZE: 6 - 10 in., max. 1 ft.

1. Dark body covered with white spots. 2. Pale honeycomb markings on body.

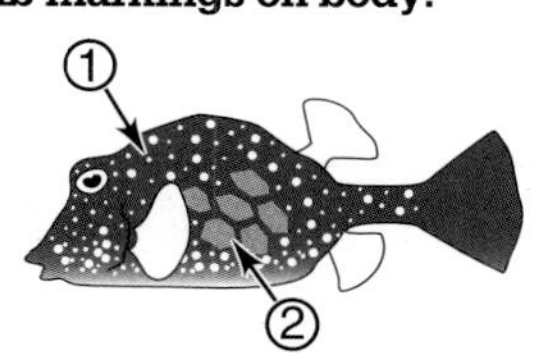

SPOTTED TRUNKFISH

Lactophrys bicaudalis

SIZE: 6 - 12 in., max. 16 in.

1. White, with black spots. 2. White around mouth.

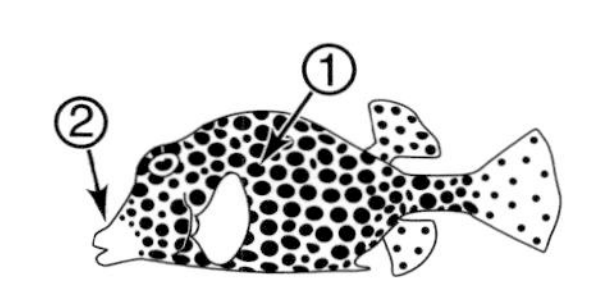

QUEEN TRIGGERFISH

Balistes vetula

SIZE: 8 - 16 in., max. 2 ft.

1. Streaming tips on rear dorsal and tail fins. 2. Lines radiate from eyes.

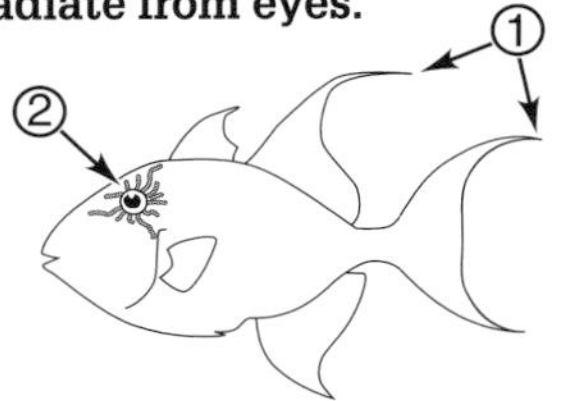

BLACK DURGON

Melichthys niger

SIZE: 6 - 12 in., max. 16 in.

1. Pale blue lines along base of dorsal and anal fins.

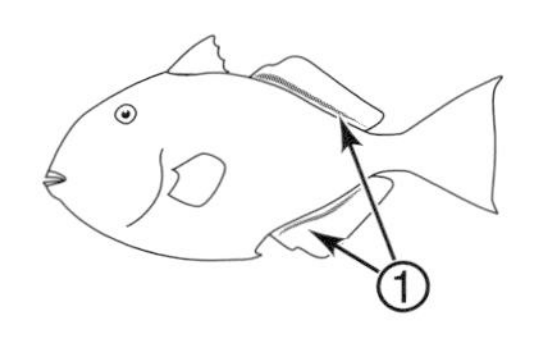

SCRAWLED FILEFISH

Aluterus scriptus

SIZE: 1 - 2½ ft., max. 3 ft.

1. Covered with blue spots. 2. Long broom-like tail.

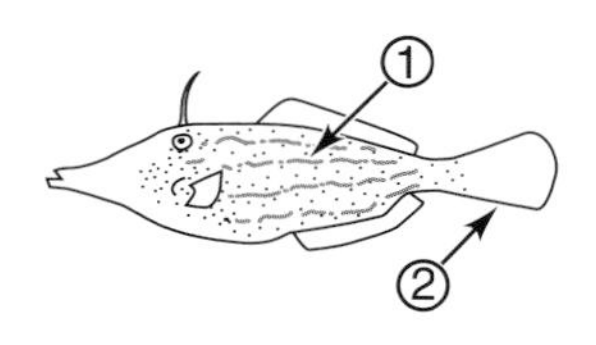

WHITESPOTTED FILEFISH

Cantherhines macrocerus

SIZE: 10 - 15 in., max. 18 in.

1. Noticeably extended belly apendage. Occasionally with large, whitish spots.

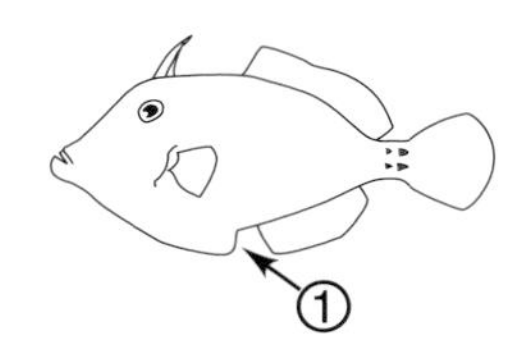

ORANGESPOTTED FILEFISH

Cantherhines pullus

SIZE: 4-7 in., max. 8 in.

1. White spot on upper tail base.

HIGHHAT

Equetus acuminatus

SIZE: 5-8 in., max. 9 in.

1. Black and white striped body.

SPOTTED DRUM

Equetus punctatus

SIZE: 6-9 in., max. 11 in.

1. Rear dorsal and tail fins black with white spots.

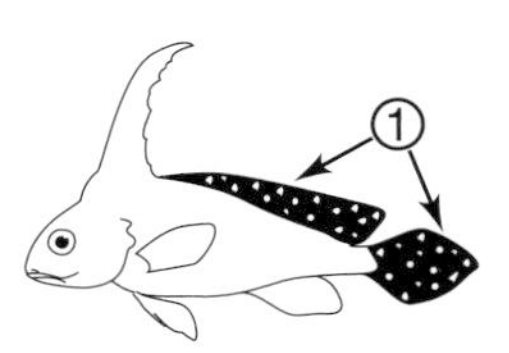

Spotted Drum Juvenile

SIZE: 3/4 - 1 1/2 in.

1. Black spot on nose.

FISH IDENTIFICATION GROUP 10

Eels, Rays & Sharks

Moray - Snake Eel - Ray - Nurse Shark

This ID Group consists of long, snake-like or flattened oval bottom-dwellers and sharks.

MORAY - Muraenidae - Morays have no pectoral or ventral fins; their dorsal, tail and anal fins form a single, long continuous fin that begins behind the head, encircles the tail and extends midway down the belly. Their heavy, scaleless bodies are coated with a clear, protective mucus layer. Morays constantly open and close their mouths, which is often perceived as a threat, but in reality this behavior is necessary to move water through their gills for respiration. They are not aggressive, although they can inflict a nasty bite if molested. During the day, they are reclusive and tend to hide in dark recesses. Occasionally, they are seen with their heads extended from holes.

SNAKE EEL - Ophichthidae - Most species of snake eels are virtually without fins and strongly resemble snakes. In fact, when first encountered, uninformed snorkelers think they are seeing a sea snake, although none inhabit the waters of Florida, the Caribbean or Bahamas. Generally snake eels are shy, reclusive fish that hide in dark recesses or burrow under sand during the day with their heads, occasionally exposed. Although they normally hunt at night, at times they are sighted foraging in the open during daylight hours.

RAY - Rajiformes - The rays' greatly enlarged pectoral fins, which give them a disc-like shape, are used for swimming, much like birds use their wings for flight. Bottom-dwelling species include stingrays, electric rays and guitarfish. Eagle Rays never rest on the bottom. Although they spend a great deal of time in deeper water they regularly feed in the shallows. Electric rays are distinguished by their circular shape and thick, short tails. As their common family name implies, these fish have the ability to discharge an electrical shock.

NURSE SHARK - Rhincodontidae - Nurse sharks are the only shark species regularly sighted by snorkelers in our region. They are usually seen resting on the bottom under the protection of ledges. They are not aggressive, but should not be molested. Other species of reef shark seldom swim over reefs by day.

GREEN MORAY

Gymnothorax funebris

SIZE: 3-5 ft., max. 8 ft.

Uniform green to brown.

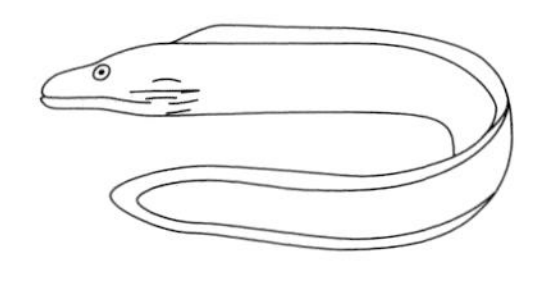

SPOTTED MORAY

Gymnothorax moringa

SIZE: 1 1/2 - 3 ft., max. 4 ft.

1. Speckling of dark spots and blotches cover body.

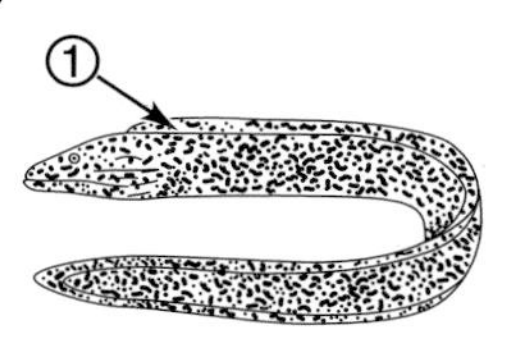

PURPLEMOUTH MORAY

Gymnothorax vicinus

SIZE: 1 1/2 - 3 ft., max. 4 ft.

1. Brilliant yellow eyes. 2. Dark edge on dorsal fin.

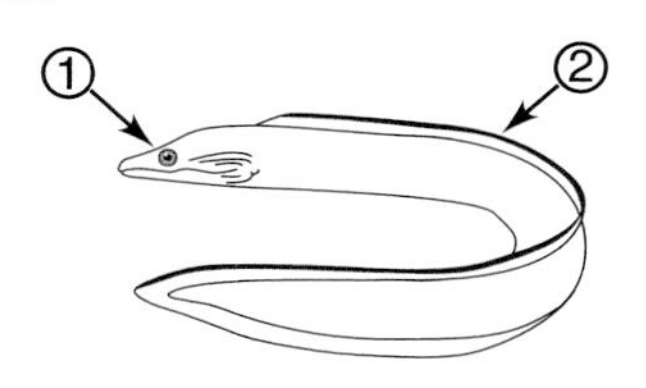

GOLDENTAIL MORAY

Gymnothorax miliaris

SIZE: 1 - 1 1/2 ft., max. 2 ft.

1. Covered with small yellow spots.

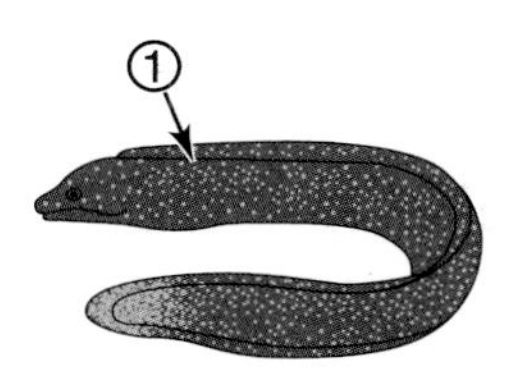

CHAIN MORAY

Echidna catenata

SIZE: 1 - 1 1/2 ft., max. 2 1/2 ft.

1. Pale to bright yellow chain-like markings.

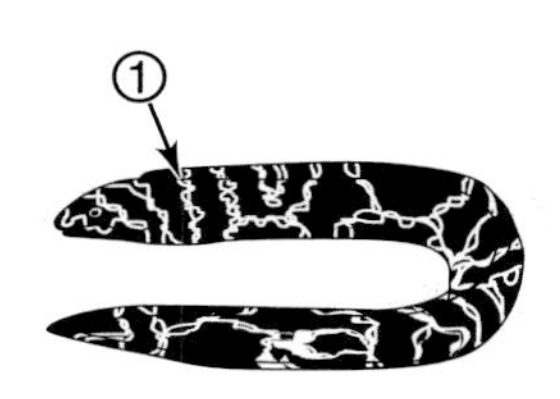

SHARPTAIL EEL

Myrichthys breviceps

SIZE: 1 - 1 1/2 ft., max. 3 1/2 ft.

1. Small yellow spots on head. 2. Pale spots on body.

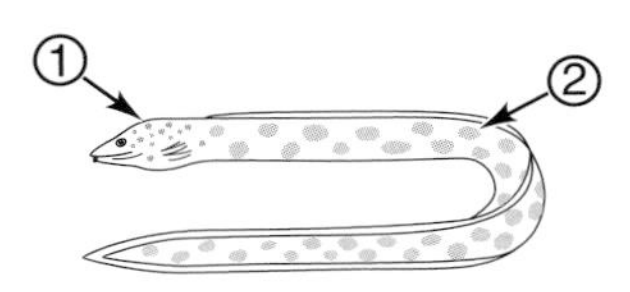

GOLDSPOTTED EEL

Myrichthys ocellatus

SIZE: 1 - 2 1/2 ft., max. 3 1/2 ft.

1. Bright gold spots with black borders on body and head.

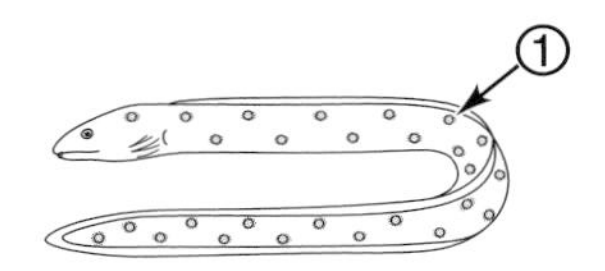

SOUTHERN STINGRAY

Dasyatis americana

SIZE: 3 - 4 ft., max. 5 1/2 ft.

1. Snout and tips of "wings" pointed.

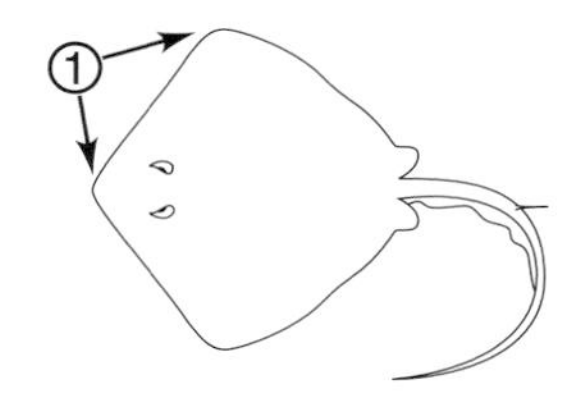

YELLOW STINGRAY

Urolophus jamaicensis

SIZE: 8 - 12 in., max. 15 in.

1. Snout and tips of "wings" rounded.

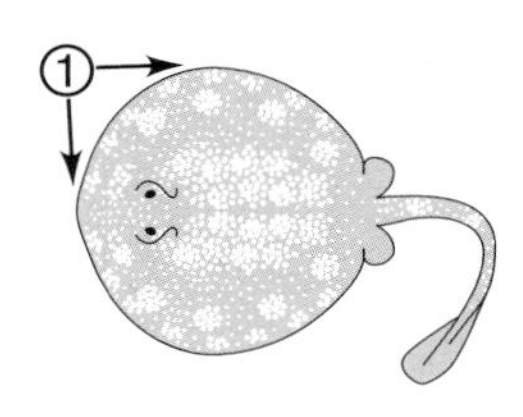

LESSER ELECTRIC RAY

Narcine brasiliensis

SIZE: 10 - 15 in., max. 18 in.

1. Two dorsal fins on tail.

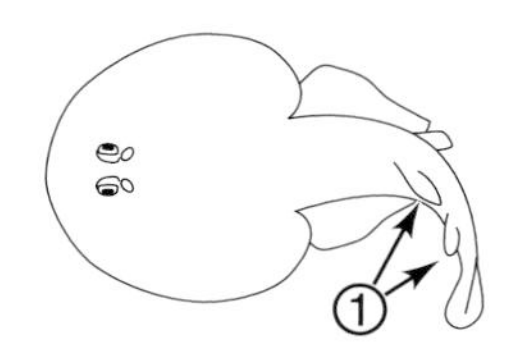

SPOTTED EAGLE RAY

Aetobatus narinari

SIZE: 4 - 6½ ft., max. 8 ft.

1. White spots over dark back. Often in shallows feeding on mollusks.

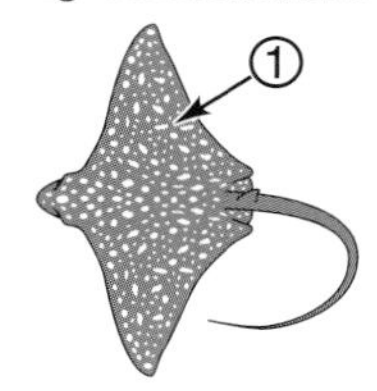

ATLANTIC GUITARFISH

Rhinobatos lentiginosus

SIZE: 1 - 2 ft., max. 2½ ft.

1. Head and pectoral fins form a triangular ray-like forebody.

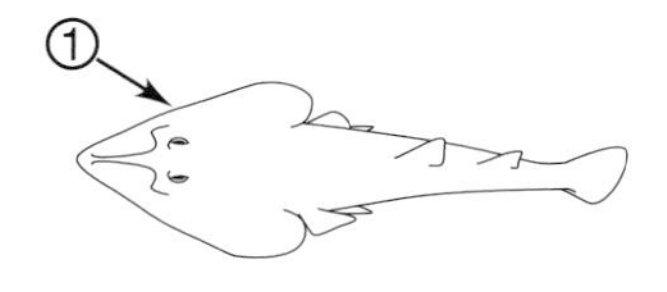

NURSE SHARK

Ginglymostoma cirratum

SIZE: 5 - 9 ft., max. 14 ft.

1. Two barbels on upper lip. 2. Two dorsal fins, of nearly equal size, are set far back.

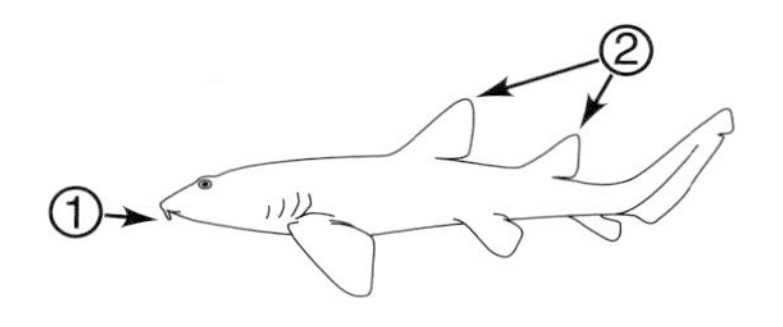

Stony Corals

Stony corals, often called hard corals, are the basic building blocks of tropical coral reefs. These animals (polyps) secrete calcium carbonate to form hard cups, called corallites, that provide protection for their soft delicate bodies. Polyps are generally smooth and tubular with tentacles in multiples of six that encircle the mouth.

The outer surface of the polyp's body has a number of vertical infolds. Calcium carbonate deposited in these folds forms thin, upright, radiating plates or ridges, called septa. The corallite structures of many species project above the overall colony forming distinctive rims, called calices. The central axis, called the columella, is below the polyp's mouth.

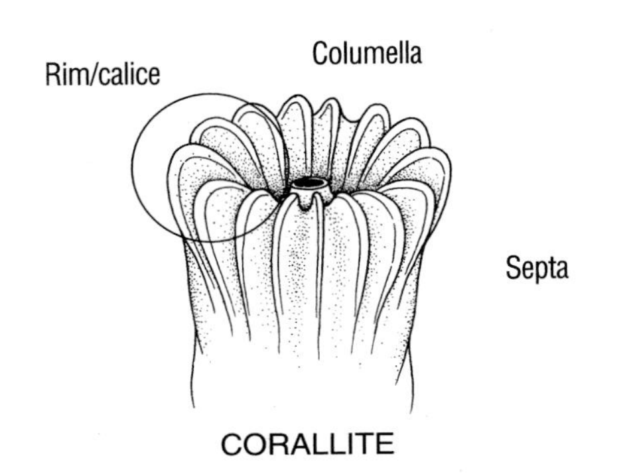

CORALLITE

In tropical waters most species grow colonially, joining their corallites to produce a substantial structure. Colonies increase in size by asexual budding of additional polyps. Successive generations overgrow one another. The size, shape and design of these structures vary from species to species. The polyps of most Caribbean stony corals are usually retracted into their corallites during the day. At night they extend both their bodies and tentacles for feeding, giving the colony a dramatically different appearance.

Colonial corals that contribute substantial amounts of calcium carbonate (limestone) to the reef structure are called hermatypic or reef-building corals. They live within a narrow temperature range, generally between 70 and 85 degrees Fahrenheit. Corals typically get their color from single-celled algae, called zooxanthellae (zo-zan-THEL-ee), that live in the polyp's tissues. This symbiotic relationship is not fully understood, but clearly the processes of each are useful to the other.

Most importantly, zooxanthellae seem to stimulate or aid the secretion of calcium carbonate. Without the algae, coral growth slows dramatically and the polyp's tissues are transparent to translucent revealing the white calcium carbonate skeleton beneath. What causes the algae to be expelled from the polyp's tissues is currently a matter of great scientific concern and debate. It is known that this process, called bleaching, takes place during times of stress; for example, after hurricanes and when water temperatures are unusually high. The current fear is that global warming is causing the abnormally high incidence of bleaching and that this may ultimately affect the diversity of corals found on reefs.

ELKHORN CORAL *Acropora palmata*

STAGHORN CORAL *Acropora cervicornis*

FINGER CORAL *Porites porites*

PILLAR CORAL *Dendrogyra cylindrus*

YELLOW PENCIL CORAL *Madracis mirabilis*

TUBE CORAL *Cladocora arbuscula*

LETTUCE CORAL *Agaricia agaricites*

LOBED STAR CORAL *Montastraea annularis*

MOUNTAINOUS STAR CORAL *Montastraea faveolata*

GREAT STAR CORAL *Montastraea cavernosa*

SYMMETRICAL BRAIN CORAL *Diplora strigosa*

KNOBBY BRAIN CORAL *Diploria clivosa*

MASSIVE STARLET CORAL *Siderastrea siderea*

BOULDER BRAIN CORAL *Colpophyllia natans*

Fire Corals

Fire coral, or stinging coral as it is sometimes called, often produces a painful burning sensation when touched by bare skin. The pain is usually short-lived and neither severe nor dangerous. For a few sensitive individuals, however, it can cause redness and welts that can last for several days. This reaction is caused by unusually powerful batteries of stinging capsules (nematocysts) located on the tentacles of the tiny polyps.

In the event of a sting, never rub the affected area or wash with fresh water or soap. Both actions can cause untriggered nematocysts to discharge. Saturating the affected area with vinegar immobilizes unspent capsules; a sprinkling of meat tenderizer may help alleviate the symptoms.

BRANCHING FIRE CORAL *Millepora alcicornis*

BLADE FIRE CORAL *Millepora complanata*

Hydroids

Hydroids are usually colonial, and have a branched skeleton that generally grows in patterns resembling feathers or ferns. Individual polyps are attached to this structure. Most species are whitish or neutral shades, ranging from brown to gray or black and rarely display vibrant colors. The stinging nematocysts of several hydroids are toxic enough to cause a painful burning sensation that may produce a visible rash, redness, or even welts.

Most hydroids have a complex life cycle. The polyps in an adult colony are specialized for either feeding or reproduction. The reproductive polyps give rise to buds that form tiny, free-swimming medusae that resemble jellyfish.

STINGING HYDROID *Macrorhynchia allmani*

ALGAE HYDROID *Thyroscyphus ramosus*

Jellyfish

Jellyfish are translucent, unattached polyps or medusas that typically swim near the surface in open water. They have prominent domes which vary in shape from a shallow saucer to a deep bell. Hanging from the margins of the domes are nematocyst-bearing tentacles, whose number and length vary greatly from species to species. The mouth is at the end of a feeding tube that extends from the center of the dome's underside. Jellyfish move through the water by pulsating contractions of the dome. Although only a few jellyfish are toxic, caution should be taken with all species.

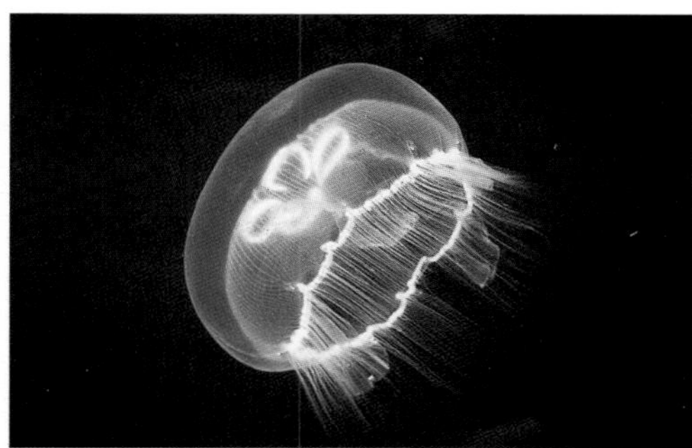

MOON JELLY *Aurelia aurita*

UPSIDEDOWN JELLY *Cassiopea frondosa*

Gorgonians

Gorgonian is the preferred name of a large group of octocorallians (polyps with eight tentacles); however, they are commonly called "soft corals" because of the colonies' lack of hard, rigid, permanent skeletons. Gorgonians include the animal colonies known as sea rods, sea whips, sea plumes, and sea fans.

The stems and branches of all species have a central skeleton or axis. The central core is composed of either tightly bound or fused calcareous spicules. The core is surrounded by gelatinous material called the rind. Polyps are embedded in the rind and extend their tentacles and bodies through surface openings called apertures. The polyp's eight tentacles bear tiny feather-like projections called pinnules. The colonies have a bushy appearance when the polyps are extended to feed. Most gorgonian colonies are attached to the substrate by a single holdfast at the base of a stem that usually branches. Branching may be in a single plane or bushy. Tiny, profuse branches called branchlets line the sides of primary branches on some species.

Telesto colonies grow by extending a long, terminal polyp that produces a stem with short side branches tipped with daughter polyps. The polyps are brilliant white. The stems are often obscured by encrusting algae, sponge and other organisms. Telesto are generally found in areas of moderate turbidity, and only rarely occur on clear water reefs. They are considered a fouling organism.

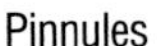

Tentacle

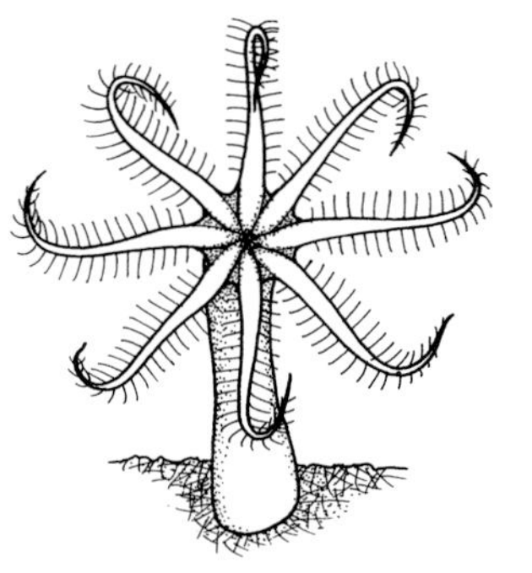

OCTOCORALLIAN POLYP

CORKY SEA FINGER *Briareum asbestinum*

BLACK SEA ROD *Plexaura homomalla*

BENT SEA ROD *Plexaura flexuosa*

POROUS SEA RODS *Pseudoplexaura sp.*

SWOLLEN-KNOB CANDELABRUM *Eunicea mammosa*

SLIT-PORE SEA RODS *Plexaurella sp.*

YELLOW SEA WHIP *Pterogorgia citrina*

GROOVED-BLADE SEA WHIP *Pterogorgia guadalupensis*

SEA PLUMES *Pseudopterogorgia sp.*

COMMON SEA FAN *Gorgonia ventalina*

VENUS SEA FAN *Gorgonia flabellum*

WHITE TELESTO *Carijoa riisei*

Sea Anemones & Zoanthids

Sea anemones are large, solitary polyps (from a few inches to a foot across) that attach to the bottom. Stinging nematocysts on the tentacles rarely affect divers, but are toxic enough to paralyze small fish and invertebrates. Several species of fish, shrimp and crab, not affected by the sting, live in association with certain anemones.

Although they rarely move, anemones can relocate in a slow, snail-like manner. The animals prefer secluded areas of the reef where they often lodge in crevices with only their tentacles exposed. If they are disturbed they can contract their tentacles for protection.

Zoanthids appear similar to anemones, but are considerably smaller, usually no larger than a half inch, and are generally colonial or live in close proximity to one another.

GIANT ANEMONE *Condylactis gigantea*

CORKSCREW ANEMONE *Bartholomea annulata*

SUN ANEMONE *Stichodactyla helianthus*

BRANCHING ANEMONE *Lebrunia danae*

MAT ZOANTHID *Zoanthus pulchellus*

WHITE ENCRUSTING ZOANTHID *Palythoa caribaeorum*

Segmented Worms

Marine worms that inhabit reefs come in a variety of shapes. Some, like free-moving fireworms, are easily distinguished while others that attach to the substrate are difficult to imagine as worms. The segmented bodies of these species are buried inside coral, rock or parchment-like tubes attached to the reef. The only part of their bodies that is visible is a crown of delicate feather-like appendages called radioles. These work both as gills, and for capturing tiny planktonic prey.

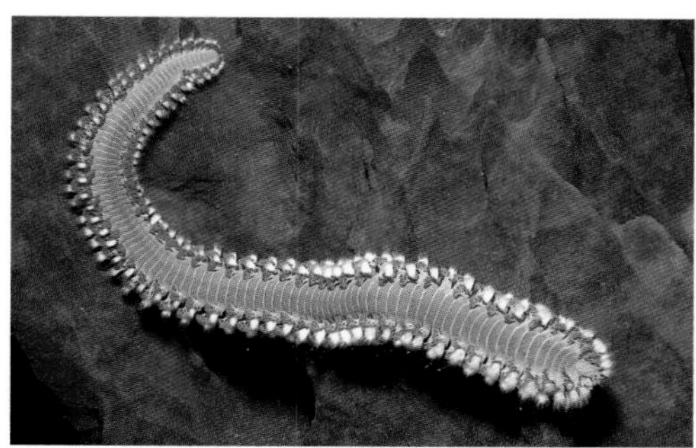

BEARDED FIREWORM *Hermodice carunculata*

SOUTHERN LUGWORM  *Arenicola cristata*

SOCIAL FEATHER DUSTER *Bispira brunnea*

CHRISTMAS TREE WORM *Spirobranchus giganteus*

STAR HORSESHOE WORM *Pomatostegus stellatus*

SPAGHETTI WORM *Eupolymnia crassicornis*

Sponges

Sponges are the simplest of the multicellular animals. The individual cells display a considerable degree of independence, and form no true tissue layer or organs. Water is drawn into the sponge through pores and pumped through the interior by the beating of whip-like extensions. As water circulates, food and oxygen are filtered out. The water exits into the body's cavity and out the animal's one or more large excurrent openings.

Sponges come in many sizes, colors and shapes. Some are quite small, less than half an inch across, while others may attain a height of over six feet. Their colors range from drab grays and browns to bright reds, oranges, yellows, greens and violets. The shape of what can be considered a typical sponge resembles a vase. However, growth patterns vary tremendously. Those with one large body opening form bowls, barrels and tubes. Sponges with multiple body openings may form irregular masses, or shapes like ropes, candles, branching horns, or, in the case of encrusting sponges, take the shapes of what they overgrow.

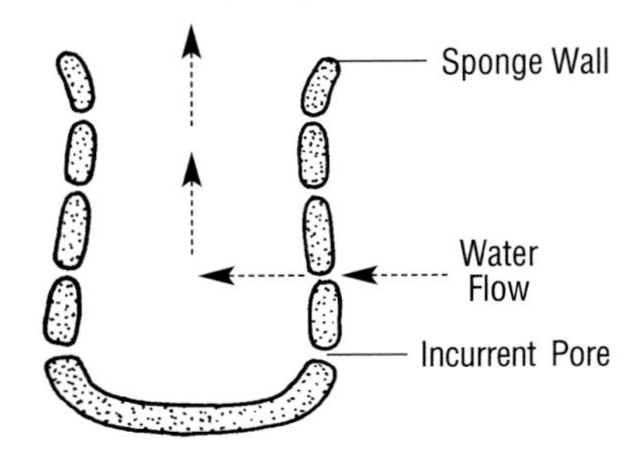

TYPICAL SPONGE CROSS SECTION

Although sponges come in many forms, they can usually be recognized as a group by their excurrent openings that are generally large and distinct. Another key is their lack of any evident movement. Nearly all animals react with an obvious protective movement when approached or touched; however, sponges show no reaction when disturbed.

YELLOW TUBE SPONGE *Aplysina fistularis*

BROWN CLUSTERED TUBE SPONGE *Agelas wiedenmyeri*

BRANCHING VASE SPONGE *Callyspongia vaginalis*

LUMPY OVERGROWING SPONGE *Holopsamma helwigi*

LOGGERHEAD SPONGE *Spheciospongia vesparium*

BLACK-BALL SPONGE *Ircinia strobilina*

STINKER SPONGE *Ircinia felix*

FIRE SPONGE *Tedania ignis*

TOUCH-ME-NOT SPONGE *Neofibularia nolitangere*

GREEN FINGER SPONGE *Iotrochota birotulata*

BROWN VARIABLE SPONGE *Anthosigmella varians*

RED BORING SPONGE *Cliona delitrix*

Crustaceans

Crustaceans are distinguished by two pairs of antennae and three distinct body parts — head, thorax and abdomen. The largest order of crustaceans is the decapods. They have five pairs of legs, and include shrimp, lobsters and crabs. The head and thorax are fused and covered by a dorsal shell called a carapace.

Shrimps' most distinguishing characteristic is their long, hair-like antennae which help snorkelers spot their hiding places inside dark recesses, within anemones, and near the openings of sponges. A number of species, called cleaning shrimp, feed by removing parasites and bacterial debris from fish.

Lobsters are bottom-dwellers that use well-developed legs to crawl about. If danger approaches, they can swim backward with darting speed, using powerful strokes of their wide, flattened tails. Snorkelers can spot spiny lobster by looking for their long , conical antennae sticking out from hiding places.

Hermit crabs use discarded sea shells as mobile homes. They occupy the shells by wrapping their long abdomen around the internal spirals of the shell. When they outgrow their homes, they simply move into a larger shell.

Crabs' claws are used for protection and for the manipulation of objects. Using the remaining four pairs of legs, crabs can move rapidly in a sideways direction. Many species are quite small, secretive, and difficult to find.

Mantis Shrimp inhabit both reefs and burrow in sandy areas. Their powerful claws can slice a finger as well as capture prey. They have an elongated body and feather-like gills attached to the lower abdomen.

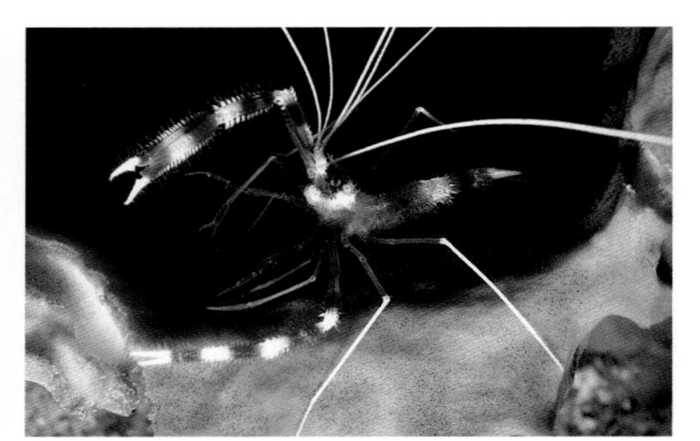

BANDED CORAL SHRIMP *Stenopus hispidus*

GOLDEN CORAL SHRIMP *Stenopus scutellatus*

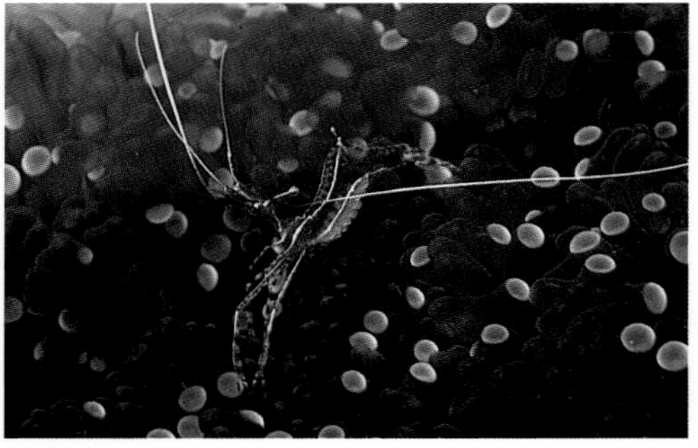

PEDERSON CLEANER SHRIMP *Periclimenes pedersoni*

RED SNAPPING SHRIMP *Alpheaus armatus*

CARIBBEAN SPINY LOBSTER *Panulirus argus*

SPOTTED SPINY LOBSTER *Panulirus guttatus*

SPANISH LOBSTER *Scyllarides aequinoctialis*

GIANT HERMIT *Petrochirus diogenes*

BATWING CORAL CRAB *Carpilius corallinus*

CHANNEL CLINGING CRAB *Mithrax spinosissimus*

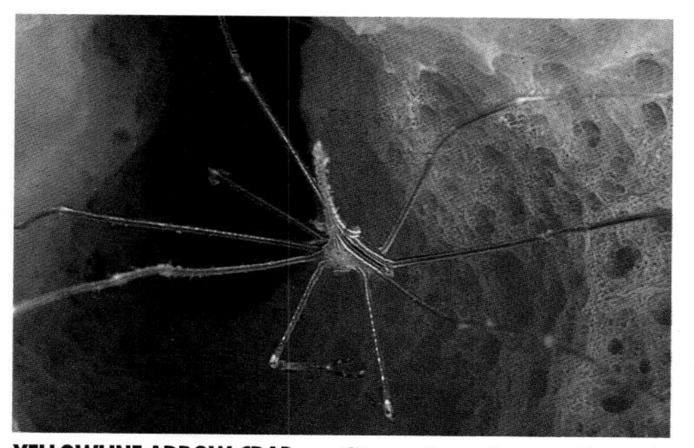

YELLOWLINE ARROW CRAB *Stenorhynchus seticornis*

SCALY-TAILED MANTIS *Lysiosquilla scabricauda*

Mollusks

Snails secrete tubular whorls that form ever-enlarging cone-shaped shells as they grow. Normally, the snails' soft bodies are completely hidden within the shells. Occasionally, a tube-like mouth, called the proboscis, and two tentacles are seen extending from the shell's opening. A short leathery pad or foot is extended to slowly drag the snail about.

Chitons are easily recognized by their oval shape, formed by eight overlapping plates. Only an inch or two in length, they attach firmly to rocks with a large, muscular foot. Movement is imperceptibly slow as the chiton grazes on algae.

Sea slugs lack an external shell. The mantle, which often has colorful, ornate designs, covers their elongated oval bodies. These ruffles increase the absorption area for oxygen which is taken in through the skin.

Bivalves are soft-bodied animals protected by two shells, called valves, that are hinged together by a ligament. When open, some species extend a curtain-like mantle that can be brightly colored.

Squid have eight arms and two longer tentacles that stream behind their elongated bodies as they swim. These fascinating creatures are often seen in groups, moving over the reef in close formation.

Octopuses have eight arms of about equal length, and globular or bag-like bodies. They are primarily bottom-dwellers that use their arms and suction cups to move about; however, they have the ability to jet backwards by rapidly expelling water from the mantle cavity. A cloud of dark ink is sometimes discharged to cover an escape.

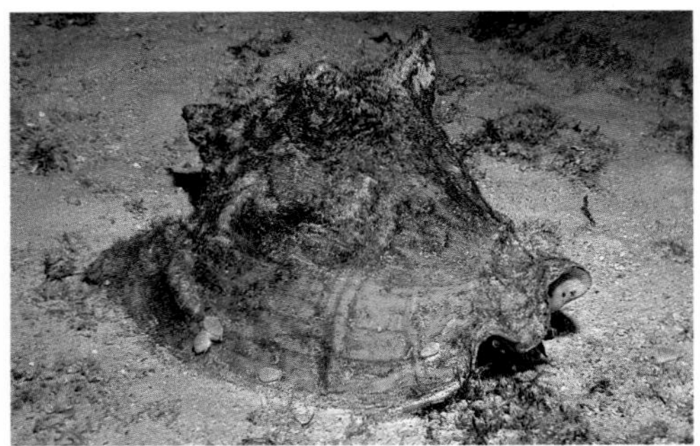

QUEEN CONCH *Strombus gigas*

MILK CONCH *Strombus costatus*

FLORIDA HORSE CONCH *Pleuroploca gigantea*

KING HELMET *Cassis tuberosa*

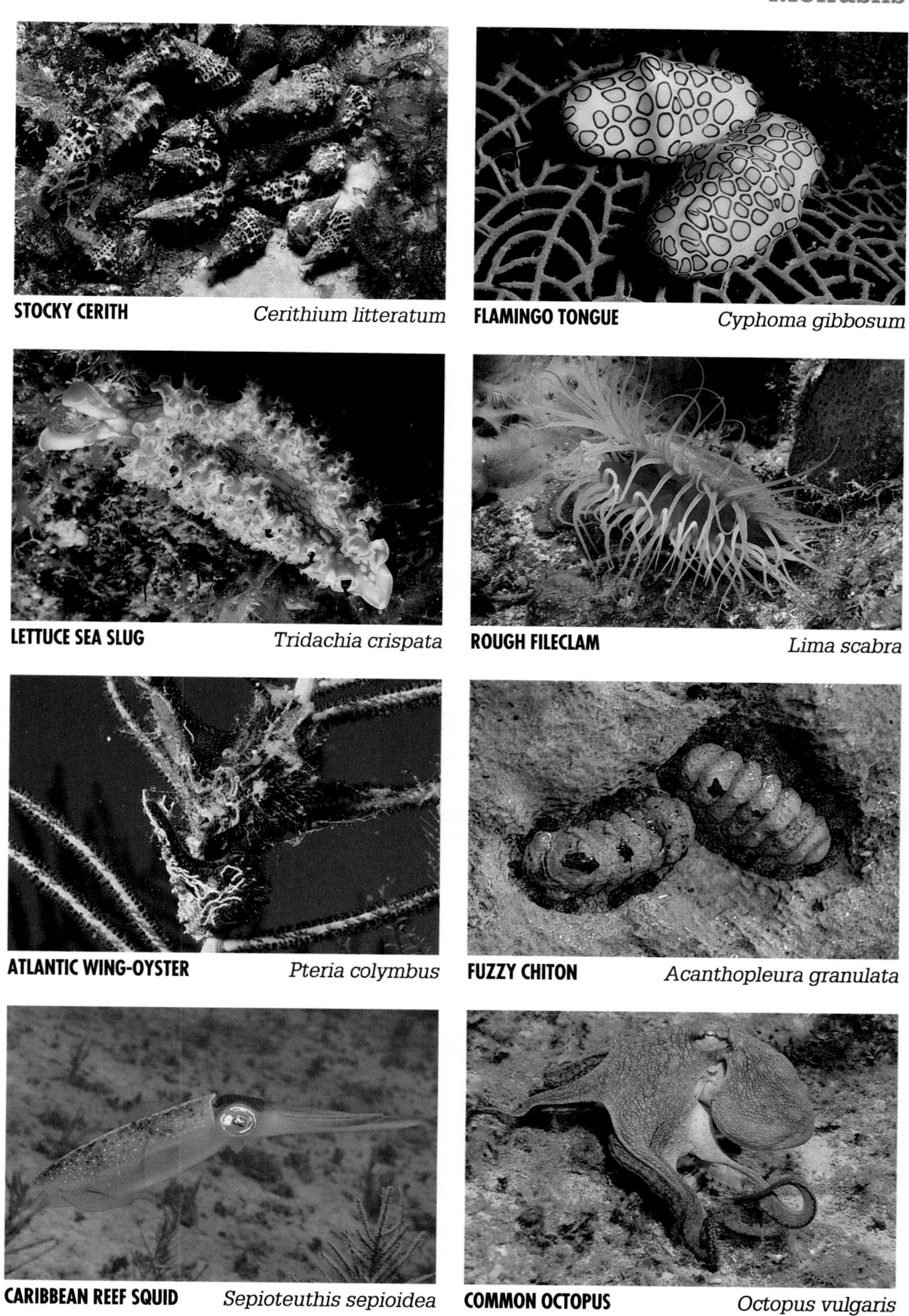

STOCKY CERITH *Cerithium litteratum*

FLAMINGO TONGUE *Cyphoma gibbosum*

LETTUCE SEA SLUG *Tridachia crispata*

ROUGH FILECLAM *Lima scabra*

ATLANTIC WING-OYSTER *Pteria colymbus*

FUZZY CHITON *Acanthopleura granulata*

CARIBBEAN REEF SQUID *Sepioteuthis sepioidea*

COMMON OCTOPUS *Octopus vulgaris*

Echinoderms

Echinoderms have five body sections of equal size that are arranged around a central axis. Most have hundreds of small tube feet that work in unison to move the creature over the bottom.

Sea stars, also commonly known as starfish, have five arms, although a few species have more. Broken arms can be regenerated, and in some species a new animal can form from a severed member. The mouth is located centrally on the undersurface, with the anus on the top.

Brittle stars have a small central disc that rarely exceeds one inch in diameter and five arms with numerous spines arranged in rows. The spines of different species can be distinct: they can be short or long, thin or thick, pointed or blunt. The tops of the arms are lined with large calcareous plates that allow only lateral movement. This armor results in arms that break off easily, giving rise to the common name. Severed arms, however, can be regenerated. During the day brittle stars are occasionally seen clinging to sponges and gorgonians, but they generally hide under rocks and inside crevices.

Sea urchins typically have spherical bodies with long protective spines and tube feet. The mouth is a complicated arrangement of five teeth, called Aristotle's Lantern, used for scraping algae and other organic food from rocks. Long, pointed spines of some species easily puncture the skin, and are difficult to remove because the shaft is covered with recurved spinelets. If left alone, they will dissolve in the tissue in a few weeks; however, the wound should be treated to prevent infection.

Heart urchins are shaped like oval domes. Their bodies are covered with short, tightly packed spines well-adapted for burrowing. Most of their lives are spent buried in sand or mud where they feed on organic material. Skeletal remains are occasionally found on the sand.

Sand dollars have a five-petal sculptured design on the back. The very short, compacted spines that cover the body appear as fuzz. Like heart urchins, they live under the sand and are rarely sighted in the open.

Sea cucumbers have sausage-shaped bodies, with a mouth in front and the anus at the rear. The five body sections common to all echinoderms are not visible in these animals, but are part of the internal structure. They also have no external spines or arms, and the skeletal plates are reduced to microscopic size and buried in the leathery body wall. Sea cucumbers are usually sighted slowly crawling across sand or reef, scooping up organic debris.

CUSHION SEA STAR *Oreaster reticulatus*

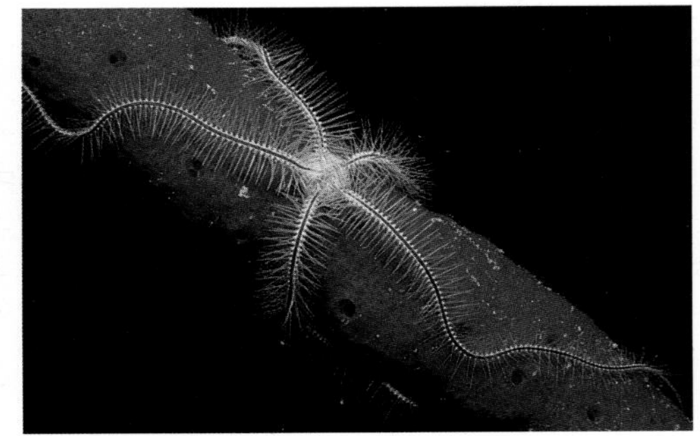

SPONGE BRITTLE STAR *Ophiothrix suensonii*

BLUNT-SPINED BRITTLE STAR *Ophiocoma echinata*

LONG-SPINED URCHIN *Diadema antillarum*

SLATE-PENCIL URCHIN *Eucidaris tribuloides*

WEST INDIAN SEA EGG *Tripneustes ventricosus*

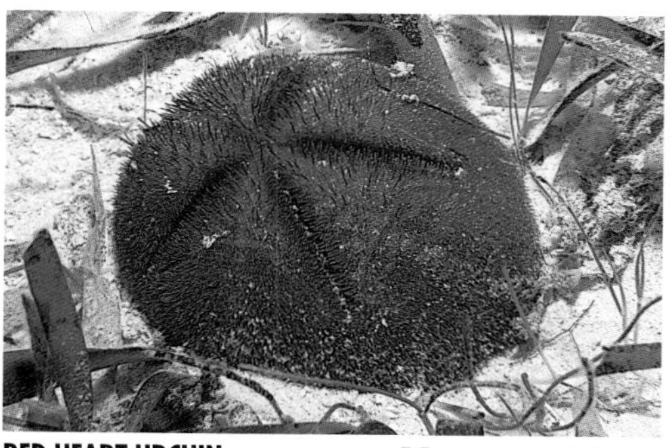
RED HEART URCHIN *Meoma ventricosa*

SAND DOLLAR *Clypeaster subdepressus*

DONKEY DUNG SEA CUCUMBER *Holothuria mexicans*

TIGER TAIL SEA CUCUMBER *Holothuria thomasi*

Marine Plants

Marine plants form the base of the oceanic food chain. All plants are photosynthetic, taking energy from sunlight and nutrients from the water or substrate, producing the food and oxygen used by other organisms to sustain life. Because marine plants require sunlight most grow in shallow clear water. There are two basic types of marine plants. Flowering plants, like their terrestrial counterparts, have true roots, stems, leaves and flowers. Algae have no true roots, but attach to the substrate by holdfast structures called rhizoids or rhizomes.

Sea grasses (not related to terrestrial grasses) often grow in great beds in areas of sand and sandy rubble. There are only two common species.

Sargassum is a brown algae. Gas-filled floats keep the plant erect or floating on the surface. Floats of Sargassum Seaweed vary from small clumps to huge rafts covering enormous areas. Many strange marine creatures make their home in these amber gardens.

Green algae are most common on tropical reefs. They are both abundant and represented by a large number of species. Many are calcareous and, along with coralline red algae, add significant amounts of calcium carbonate which creates much of the brilliant white sand found on and around reefs. Green algae grow in a wide range of patterns, including single-celled bubble-like plants to fan-like blades, bristly brushes, hanging vines and serrated-edged, upright blades. Chlorophyll is the pigment primarily responsible for the green and yellowish green color.

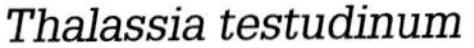

TURTLE GRASS *Thalassia testudinum*

MANATEE GRASS *Syringodium filiforme*

SARGASSUM SEAWEED *Sargassum fluitans*

SARGASSUM ALGAE *Sargassum sp.*

Y BRANCHED ALGAE *Dictyota sp.*

SERRATED STRAP ALGA *Dictyota ciliolata*

LEAFY FLAT-BLADE ALGA *Stypopodium zonale*

LEAFY ROLLED-BLADE ALGA *Padina boergesenii*

WHITE SCROLL ALGA *Padina jamaicensis*

WATERCRESS ALGA *Halimeda opuntia*

FLAT-TOP BRISTLE BRUSH *Penicillus pyriformis*

GREEN FEATHER ALGA *Caulerpa sertularioides*

GREEN GRAPE ALGA *Caulerpa racemosa*

MERMAID'S FANS *Udotea sp.*

PINECONE ALGA *Rhipocephalus phoenix*

GREEN MERMAID'S WINE GLASS *Acetabularia calyculus*